BIBLIOTHECA PHYCOLOGICA

Herausgegeben von
J. CRAMER

BAND 54

Contribution to the Marine Algae of Libya Dictyotales

by

MOHAMMED NIZAMUDDIN

with 39 plates

1981 · J. CRAMER
In der A.R. Gantner Verlag Kommanditgesellschaft
FL-9490 VADUZ

Authors address:
Department of Botany
University El-Faateh, Tripoli
Libya

© 1981 A.R. Gantner Verlag K.G., FL-9490 Vaduz
Printed in Germany
by Strauss & Cramer GmbH, 6945 Hirschberg 2
ISBN 3-7682-1305-6

Dedicated to

Alexander von Humboldt-Stiftung
Bad Godesberg-Bonn, West Germany

to which I am grateful
for making it possible to continue
my studies on Libyan algae

INDEX

INTRODUCTION

Libya occupies a large part of the north African coast of the Mediterranean Sea and is mostly sandy but there are some rocky parts belonging to the Pleistocene as well as some to Miocene cliffs. There are also sharp rocks, near the sea-level, due to strong cementation and pinnacle erosion by boring and grazing organisms or by inorganic solutions. Some hard crust rocks are also found here and there along the coast.

Distribution of surface salinity ranges from 38-40°/₀₀ and the tidal level ranges from 0-1 m. There are great fluctuations in air temperatures also. The Oceanographic and the Hydrographic data are lacking from this coast. There has never been any organised scientific effort to exploit the marine algal flora of Libyan coast. During the Italian's reign few and scattered algal collections were made, identified and published by De Toni et Levi (1888), De Toni et Forti (1913, 1914) and Piccone (1892). During the Danish Oceanographic Expedition 1908-1910 some algal collections were made by dredging and these were identified by Petersen (1918). Synoptical lists of algae were published by Huschler (1910) and Pampanini (1931) based on the above authors' contributions.* Earlier phycological studies were based on few and fragmentary specimens which were deposited either in Florence or Copenhagen. Presently efforts were or are being made to collect marine algae from all along the coast of Libya. Author's account is the first attempt to study the marine algae in detail from the Libyan coast. The present study deals with the morpho-taxonomy and ecology of the members of the order Dictyotales. In addition to the Dictyotales of the Libyan coast the author has also studied the Dictyotales from the European coast

_____

Nizamuddin, West and Menez (1979) published a comprehensive synoptical list of marine algae (excl. diatoms) based on recent nomenclatural and taxonomical aspects.

of the Mediterranean Sea and is incorporated in the text. The specimens in other herbaria which do not bear the date of collections or the name of the collectors were also examined and mentioned in the text.

MATERIAL AND METHODS

Materials were either collected as living or drift or in some cases by diving. These were fixed in formalin-ethyl alcohol-seawater (1:1:7) solutions for anatomical studies and some were mounted on herbarium sheets. In most cases freehand sections were cut either from the fresh or preserved materials. Specimens were kept in the Herbarium, Botany Department, Faculty of Science, University El-Faateh, Tripoli and also in the Botanisches Museum, Berlin-Dahlem.

DICTYOTALES

Plants thalloid, isomorphic alternation of generations; apical growth by a single convex apical cell or by marginal initials; asexual reproduction by tetra- or octa-spores; sexual reproduction by antherozoids and eggs.

DICTYOTACEAE

Thalli or fronds of moderate to large size, simple or branched, growth by an apical cell or marginal initials; fan-like, strap-shaped, foliaceous, branched or lobed segments; composed of central layer or layers of large cells and a single layer of small peripheral cells on either side of the surface; isomorphic alternation of generations; reproductive organs in sori; sporangia producing 4 or 8 spores; eggs born singly in the superficial oogonia; antheridia superficial, plurilocular, producing biflagellate antherozoids (single rudimentary flagellum seen only in electron microscope). The family comprises only seven genera which are found along the coast of Libya and the same number of genera found along the other parts of the Mediterranean coast.

10

1. Fronds with distinct midrib . . . . *Dictyopteris*
1. Fronds without midribs 2

2. Fronds fan-shaped; apical margin revolute . . . . *Padina*
2. Fronds otherwise 3

3. Fronds zonate 4
3. Fronds not zonate 5

4. Sporangia with 8 spores . . . . *Zonaria*
4. Sporangia with 4 spores . . . . *Taonia*

5. Growth by a single apical cell 6
5. Growth by marginal initials . . . . *Spatoglossum*

6. Holdfast discoid; fronds composed of a single layer of large central
   cells throughout . . . . *Dictyota*
6. Holdfast stoloniferous; fronds composed of 2 or more layers of large
   central cells in the lower or middle parts, in some cases throughout
   except apical region . . . . *Dilophus*

DICTYOPTERIS Lamouroux 1809b

Fronds small to large with velvety rhizoidal holdfast, sub-
dichotomously or laterally branched into strap-shaped segment
with percurrent midribs, sometimes with lateral veins; vegetative growth by means of marginal initials; sori or crypts scattered or linear or transversely obliquely arranged on either
side of the midribs; sori, producing reproductive organs, scattered or regularly dispersed on either side of the midrib.

Type: *Dictyopteris polypodioides* Lamouroux.

There has always been confusion and misunderstanding among the
phycologists over the nomenclature of the type species of *Dictyopteris* Lamx. To reach any definite conclusion one has to
consider *Fucus membranaceus* Stackhouse (1795), *Fucus polypo-*

11

dioides Desfontaines (1798), *Fucus membranaceus* Stackhouse of
Turner (1802, 1809), *Fucus polypodioides* Lamouroux (1805), *Neurocarpus membranaceus* Weber et Mohr (1805), *Ulva polypodioides*
de Candolle (in Lamarck et de Candolle, 1805; 1806), *Dictyopteris polypodioides* Lamouroux (1809b), *Dictyopteris elongata* Lamouroux (1809b) and *Haliseris polypodioides* C. Agardh (1820).
For ready reference the original descriptions of the above species are given below to come to a definite decision about the
type species of the genus.

Fucus membranaceus Stackhouse (1795:13) - Fronde dichotoma,
membranacea pellucida; costata ramulis, et foliolis sparsim e
costa erumpentibus. Radix plana, orbiculata, agglutinata. Cau-
lis dichotomus, cartilagenus, subnudus. Folia membranacea,
dichotoma; caule medium folii percurrente, prolifero. Fructi-
ficatio in punctis glomeratim dispositis.

Fucus membranaceus of Turner (1802:141) - Fronde lineari, dicho
toma, membranacea, punctata, diaphana, tenerrima, nerve undula-
to, sparsim prolifero.

Fucus membranaceus of Turner (1809:41) - Fronde membranacea,
plana, costata, lineari, dichotoma, integerrima, e costa prolifera; seminibus audis, in maculas hemisphaericae utrinque juxta
costam congestis.

Fucus polypodioides Desfontaines (1798:421) - Fronde dichotoma
caule folium medium percurrente; punctis tuberculosis, distinc-
tis, sparsis, subrotundis.

Fucus polypodioides Lamouroux (1805:32) - Fronde plana, stipite
mediam illam percurrente, dichotoma vel ramosa; sparsis in utr
que pagina frondis fructificationibus ad stipitem quam in margine frequentioribus.

Ulva polypodioides de Candolle (in Lamarck et de Candolle,
1805:15) - Cette espèce n'a que 1-2 décim. de longueur; elle
adhère au sol par un disque orbiculaire, d'où sortent une ou
plusieurs tiges grêles noirâtres et nues vers leur base, bien-
tôt bordées d'une membrane verdâtre, pellucide, entière en son

bord, obtuse à son sommet; cette feuille se bifurque plusieurs fois; on y remarque de petite points noire qui, vus à la loupe, sont des amas de graines séparées les unes autres. Cette plante differe d'ulve bifurquée, à cause de la côte saillante qui occupe le milieu de la feuille. Elle croit dans l'Ocean et dans la Méditerranée.

Ulva polypodioides de Candolle (in Lamarck et de Candolle, 1806: 5) - Fronde dichotoma, segmentis margine integris obtusis, punctis fructificationis sparsis.

Neurocarpus membranacea Weber et Mohr (1805:300) - Semina arillata in punctis semirotundis ad utrumque venae frondis latus, alternatum in utraque pagina, congesta, nuda.

Dictyopteris polypodioides Lamouroux (1809b:131) - Fronde membranacea ramosa; fructificationibus ad stipitem frequentibus.

Dictyopteris elongata Lamouroux (1809b:130) - Fronde membranacea ramosa elongata, tenera; fructificationibus minutis, numerosis, sparsis.

Haliseris polypodioides C. Agardh (1820:142) - Fronde lineari dichotoma integerrima, soris ad costam coacervatis.

It appears from the descriptions, plates, figures of the above species that the previous authors were dealing with two taxa not one. The description of the species *Fucus membranaceus* Stackh. does not agree with its Plate No. VI. The description states regularly arranged fructifications whereas the plate shows scattered fructifications. Turner (1802) states scattered fructifications which agree with Plate No. VI of Stackhouse but later in 1809 he described hemispherical spots of fructifications near the midrib which agree with the description of the species given by Stackhouse. Turner's description agrees with his plate No. 87. It is clear that both authors, Stackhouse and Turner, gave contradictory description with regard to the plate at one stage or other. In the absence of any type material it is practically impossible to decide whether Plate No. VI of Stackhouse or the description of his species is correct as a

13

rule both should agree but it is not so. At present type material of *F. mambranaceus* Stackh. is not traceable. Now the question is that whether the material existed or not. That material did at one time exist is hardly in question. The specimen concerned was clearly seen by Turner at some stage or other prior to his publication in 1802 of Syn. Brit. Fuci. In the original pre-annotation text (Turner's own annotated copy aimed at preparing for his Fuci . . . .), on p.p. 143-144, Turner states ". . . . I believe that only two native specimens of *F. membranaceus* are known to exist in England; one in the possession of Sir Thomas Frankland, Bart. the other in that of its first discover, Mr. Stackhouse; which latter, through his kindness, is now before me; and from it, assisted by some smaller ones . . . .". It is implied that either Turner retained the specimen that he obviously at one time saw, or that it was returned to Mr. Stackhouse. The type material derived from Sidmouth, Devon, west promontory (Leg. Mr. Stackhouse, 1794), is not traceable in BM. In the absence of the type material the illustration Plate No. VI is selected as Lectotype. Desfontaines described *F. polypodioides* having scattered fructifications but without any plate or figure. At present the type specimen is not traceable/available. Lamouroux (1805) described *F. polypodioides* which possesses scattered fructifications. His description is in agreement with his plate no. XXIV which also show scattered fructifications. Both species, *F. polypodioides* Desfont. and *F. polypodioides* Lamx., are later homonyms of Gmelin's *F. polypodioides* (1768). Under the International Code of Botanical Nomenclature *F. polypodioides* Desfont. and *F. polypodioides* Lamx. are rejected.

De Candolle (in *et al.*) made a new combination, *Ulva polypodiodes* having scattered fructifications. The description of his species is based on specimen nos. 374/4, 374/5A, 374/5B in G which possess linear arrangement of fructifictions along the midrib but not scattered as described by him. There are other specimens in his herbarium which do possess scattered fructifi cations. *Neurocarpus* Weber et Mohr (1805) is '*nomen rejiciendum*' considered by the compiler's of Nomina Conservanda, Inter national Code of Botanical Nomenclature (de Wildeman, 1910, Lanjouw, 1952, Stafleu 1972; 1978), though it has the priority

14

Dictyopteris Lamouroux (1809b) and *Haliseris* C. Agardh (1820).

Lamouroux (1809b) established a new genus, *Dictyopteris*, with its four species i.e. *D. justii*, *D. elongata*, *D. polypodioides*, *D. delicatula* without typification. He separated *D. elongata* and *D. polypodioides* on the geographical ground as well as on the arrangement of fructifications but was not justified in considering *U. polypodioides* DC (in *et al.*) and *F. polypodioides* Desfont. in separate groups though both the species possess scattered fructifications. Of course the previous authors were dealing with two distinct species. Therefore Lamouroux was correct in creating two separate species, *D. elongata* and *D. polypodioides*. In the arrangement of fructifications *D. elongata* is similar to the Plate VI of *F. membranaceus* Stackh. Stackhouse has the priority over Lamouroux. Therefore the correct legal name should be *Dictyopteris membranacea* (Stackh.) Batters. The specimens possessing fructifications in linear row along the midrib should be known as *Dictyopteris polypodioides* Lamouroux. C. Agardh (1820) described *Haliseris polypodioides* possessing sori along the midrib and considered *F. membranaceus* Stackh. and *F. polypodioides* Lamx. as synonyms. The consideration of the last two species as synonyms by C. Agardh is incorrect as both species possess scattered sori. Further more C. Agardh (l.c.) considered *F. polypodioides* Desfont. and *Dictyopteris polypodioides* Lamx. as synonyms of *H. polypodioides* var. *minor* (Lamx.) C. Ag. This consideration is also incorrect as these two species also differ from each other in the arrangement of sori.

Compiler's of Nomina Conservanda, International Code of Botanical Nomenclature (Stafleu, 1972; 1978) conserved *Dictyopteris* Lamouroux (1809b): *Dictyopteris polypodioides* (de Candolle) Lamouroux. This citation is incorrect as de Candolle did not publish "Flore francaise" separately.. His account was published in Lamarck and de Candolle (1805). So the correct citation should have been: de Candolle (in Lamarck et de Candolle 1805) Lamouroux. Even the dating of *Dictyopteris* back to 1805 by the compiler's of Nomina Conservanda, International Code of Botanical Nomenclature (de Wildeman, 1910) is incorrect as this genus was published by Lamouroux in 1809b and not in 1805.

Under the Code *Neurocarpus* (Weber et Mohr, 1805) have the priority over *Dictyopteris* (Lamouroux, 1809b) and *Haliseris* (C. Agardh, 1820) but it has already been considered 'nomen rejiciendum'. If the tradition of common use of the genus has to be followed then *Haliseris* C. Agardh should have been selected for conservation as it was more commonly used than the others.

D. polypodioides Lamouroux (1809b: 131) is considered as the type of the genus, not *D. polypodioides* (de Candolle) Lamouroux. The arrangement of sori on the segments of the fronds appears to be a constant character in *Dictyopteris* species found along the Mediterranean coast. On the basis of these features three distinct species have been recognised, *D. membranacea* (Stackh.) Batters; *D. tripolitana* sp. nov. and *D. polypodioides* Lamouroux (Plate XXXIX).

KEY TO THE SPECIES

1. Sori scattered on either side of the midrib . . . . *D. membranacea*
1. Sori arranged otherwise on the segments 2

2. Sori transversely obliquely arranged on either side of the midrib . . .
. . . . *D. tripolitana*
2. Sori linearly arranged on either side of the midrib . . *D. polypodioides*

DICTYOPTERIS MEMBRANACEA (Stackhouse) Batters
Plate I A, B; II A; XXI. Figs. A, B, C.

Basionym: *Fucus membranaceus* Stackhouse 1795: Pl. VI (Excl. desc.)

Synonyms: *Fucus membranaceus* Turner 1802: 141. *Fucus polypodioides* Desfontaines 1798: 421 non Gmelin 1768: 186. *Fucus polypodioides* Lamouroux 1805: 32, non Gmelin 1768: 186. *Ulva polypodioides* de Candolle (in Lamarck et de Candolle) 1805: 15; 1806: 5. *Dictyopteris polypodioides* (Desfont.) Lamouroux Bornet 1892: Feldmann 1931: 217. Funk 1955: 51. Hauck 1885: 311. Le Jolis 1863: 99. Reinke 1878: 36. Trevisan 1849: 456. *Zonaria polypodioides* (Lamx.) C. Agardh 1817: XXI. *Dictyopteris elongata* Lamouroux 1809b: 130; 1809c: 332; 1813: 277. Gray 1821: 342.

16

Ardrê 1969: 399. Batters 1902: 54. Feldmann 1937: 319. Gayral 1958: 228;
1966: 265. Gerloff und Geissler 1971: 750. Giaccone 1968: 223; 1969: 500.
Hamel 1939: 341. Levring 1974: 33. Newton 1931: 216. Nizamuddin and Lehn-
ber 1970: 112- Rodrigues 1963: 44. Taylor 1960: 227.

Fronds erect tufted arising from a discoid velvety holdfast,
reaching to 18 cm high and lower parts laterally or alternately
branched and thick. Upper parts subdichotomously branched. Segments flat, membranaceous, leafy to 1 cm broad with percurrent
midrib to 1 mm broad, 2-4 mm thick; margin entire, wavy, sometimes lacerated; apices broadly acute; dichotomy distinct near
the apices but in other parts sub- or trichotomous; lower parts
denuded of the wing and the midrib functioning as stipe. Crypts
in a single row near the apices but scattered in other parts.
Sori containing spores scattered on either side of the midrib.
Each sporangia producing 4 spores.

Type: *Fucus membranaceus* Stackh. Pl. VI.

Type locality: Sidmouth, Devon West Promontory (Leg. Mr. Stackhouse, 1794).

Specimens examined: CRETE - Chersónisos (Leg. J. Gerloff, 29.5.
1974, Berol. No. 29350).
GREECE - Attika, near Cape Sounion (Leg. J. Gerloff, 7.4.1971,
Berol. No. 26950, growing on rocks).
YUGOSLAVIA - Ragusa (April 1884, Berol. No. 10714). Spalato
(Leg. V. Schiffner, 16.4.1928, Berol. No. 10719). Val di Bora,
Rovinj (Leg. V. Schiffner, 17.7.1914, Berol. No. 10707).
ITALY - Trieste, port area (Leg. J. Schiller, 1-3 April 1925,
Berol. No. 24692 from 4 m depth; 15-17 June 1924, Berol. No.
10717. Leg. V. Schiffner, June 1905, Berol. Nos. 10718, 10730.
Leg. F. Krasser, Krypt. Exic. No. 1511, Berol. No. 10729). Gulf
of Trieste (Leg. V. Schiffner, 12.8.1914, Berol. No. 24680).
St. Andrea, Trieste (Leg. V. Engelhardt, 19.1.1882, Berol. Nos.
24684-5). Pegli near Genova (Ex herb. G. Zeller, Berol. Nos.
10720, 10722, 10723, 10733). Barcola (Leg. G. Seefelder, 11.12
1912, Berol. No. 10703).
FRANCE - Cap du Troc (Leg. Runze, 6.9.1962, Berol. No. 17928).
Banyuls towards Cerbére (Leg. J. Kohlmeyer, 5.6.1962, Berol.

No. 17521). Port Vendres (Leg. J. Kohlmeyer, 20.6.1962, Berol. No. 17520). Biaritz (Leg. J. Gerloff, 22.9.1961, Berol. No. 04952). Roscoff (Leg. Runze, 21.4.1962, Berol. No. 17795).

Dictyopteris membranacea is a very variable species. Schiffner mentioned many forms and varieties belonging to this species. All these names are herbarium names so these are considered *nom. nud.*, because of its polymorphic nature it is not worth to name forms or varieties unless these are based on constant characters.

Nomenclature and taxonomical aspects of this species have already been discussed in the text.

DICTYOPTERIS TRIPOLITANA sp. nov.
Plate II B, III A, XXII, XIII. Figs. a, b, c.

Diagnosis: Planta frondibus erectis, flexuosis, alternatim vel subdichotoma ramosis; segmentis membranaceis integerrimis. Costa distincta folium percurrens, proliferationibus numerosis, sori minuti, transverse-obliqui in maculas utrinque juxta costam dispositis.

Holotypus: 0512a, M. Nizamuddin.

Locus typi: Sciare El-Faateh, ad Tripoli (leg. M. Nizamuddin 15.6.1974).

Holdfast discoid having velvety fibrous rhizoids. Fronds erect caespitose reaching to 30 cm high, laterally or alternately or sub-dichotomously branched near the base; upper parts sub-dichotomously branched into leafy segments with percurrent midribs. Basal parts denuded of the wings, midribs functioning as stipes. Stipe terete to 1 cm high, to 2 mm broad, to 900 µm thick. Leafy segments laterally or sub-dichotomously branched, to 19 cm long, flat, membranaceous, 7-12 mm broad (7 mm broad near the basal parts, 12 mm broad near the upper parts), 240-850 µm thick, linear-ovate with percurrent midribs; margin entire and wavy; apices acute; midrib to 2 mm braod reaching to the apices; angle of dichotomy acute becoming divergent with age. Old or mature segments wavy or tortuous as well as lacer-

18

ated. Proliferations numerous from the midribs.

Young segments tereto-compressed to 70 μm thick near the midrib, composed of few layers of cells; wings becoming 3-4 layered cells containing dense chromatophores. Stalk of the young
segments 300-350 μm thick, composed of many layers of cells.

Mature segments composed of many layered cells near the midrib,
240-850 μm thick; wing composed of 4 layers of cells, to 120 μm
thick, becoming 2 layered cells near the apices.

Sori or crypts minute, transversely obliquely arranged on either side of the midrib.

Holotype: 0512a.

Type locality: Sciare El-Faateh, near Fishermen's Mosque,
Tripoli (Leg. M. Nizamuddin, 15.6.1974). Drift as epiphyte on
Posidonia oceanica (L.) Delile.

Specimens examined: MOROCCO - Tanger (Algae Schousboeanae, 1815;
1826 in BM). Ceuta (Leg. M. Gandogerr, 1906 in BM).
ALGERIA - Alger, near Oneel Babel (April 1840 in BM).
TUNISIA - Tabarka (Leg. J. Gerloff, 11.4.1968, Berol. No.
24523). Ain-Oktor (Leg. J. Gerloff, 10.4.1968, Berol. No.
24512). Hammamet (Leg. J. Gerloff, 13.4.1968, Berol. No. 24518).
Soliman (Leg. J. Gerloff, 19.9.1967, Berol. No. 24169). Gulf of
Gabez (Leg. J. Gerloff, 11.8.1969, 25689).
LIBYA - Zanzoor (Leg. Gadeh, 10.7.1973). Tripoli, near Gargaresh
(Leg. M. Nizamuddin, 17.4.1973). Tripoli, near Govt. Guest House
(Leg M. Nizamuddin, 12.5.1974). Tripoli, near Radio Station
(Leg. M. Gadeh, 8.7.1973); Tripoli, near Fishermen's Mosque,
Sciare El-Faateh (Leg. M. Nizamuddin, 15.6.1974); Tripoli, near
Ender Hospital (Leg. M. Nizamuddin, 10.4.1972, No. 0110; 27.9.
1974); Tripoli (cf. Muschler, 1910). El-Khoms (Leg. M. Niza-
muddin, 23.5.1974). Bengazi (Leg. M. Nizamuddin, 3.3.1973; Leg.
A.A. Razag, 18.5.1973). Bengazi-Guilliana (Leg. Zanon, 1915).
Cyrenaica (Leg. Hainmann, 1881).
MALTA - Valeta Habour (. . . May 1845 in BM).
GREECE - Attika, near Nea Makri (Leg. J. Gerloff, 11.4.1971,

Berol. No. 26953). Cape Sounion (Leg. J. Gerloff, 20.5.1972,
Berol. No. 28812). Egina Is. (Leg. J. Gerloff, 16.4.1971, Berol.
No. 27001). Paleokastritsa, Corfu (Leg. Kuhbier, May 1968,
Berol. No. 24601, from 10 m depth). Killini, Peloponisos (Leg.
J. Gerloff, 14.5.1975, Berol. No. 20718).
CRETE - North coast near Candia (Leg Watzl, 24.4.1914, Berol.
No. 24686). Chersónisos (Leg. J. Gerloff, 24.5.1974; Berol. No.
29386; 29.5.1974, Berol. No. 29361.
YUGOSLAVIA - Ragusa (Leg. V. Schiffner, 5.4.1926, Berol. No.
24688. Leg. Zay, Berol. No. 10711). Lissa Kut (Leg. J. Schil-
ler, 3.3.1912, Berol. No. 24689). Val di Bora, Rovinj (Leg. A.
Vatova, 7.3.1932, Berol. No. 10712, from 5 m depth). Rovinj,
near port (Leg. J. Gerloff, 15.6.1961, Berol. Nos. 19905-13,
from 2 m depth). Rovinj (Leg. J. Gerloff, 10.6.1961, Berol.
Nos. 19928, 26180; May 1970, Berol. No. 26312). Pirano (Leg.
Liechtenstein, Berol. No. 10716).
ITALY - Barcola (Leg. G. Seefelder, 11.12.1912, Berol. No.
10713, from 2-3 m depth). Porto Savona (Leg. A. Piccone, 1860,
Berol. No. 10724). Port Lessina (. . . April 1884, Berol. No.
24679). Casamicciola, Ischia Is. (. . . 28th April 1870, Berol.
No. 10732). Port Rende, Gulf of Naples (Leg. J. Brunnthaler,
7.4.1914, Berol. No. 10709).
SICILY - Arenella (Leg. G. Langenbach, 16.6.1870, Berol. No.
10715). Palermo (Leg. L. Kny, 15.1.1870, Berol. No. 24627; 2.2.
1870, Berol. Nos 10725B, 10727A). Aspara (Leg. G. Langenbach,
9.5.1870, Berol. No. 1072B).
ENGLAND - Sadrane Bay (Leg. Homes, 1.1.1880 in BM). Torquay
(Leg. E.H. Boning, August 1883, in BM. Leg. E. George, 30.8.
1896, in BM). Exmouth, Devon (Leg. C.A. Johnson, June 1864, in
BM). Portlet Bay, Jersey (Leg. C.A. Johnson, June 1864, in BM).
Devonshire (Leg. Mrs. Griffiths, 1806, in BM).
MALLORCA - Magaluf (Leg. J. Gerloff, 17.2.1970, Berol. No.
25832).

D. tripolitana Grows in calm mid-littoral rocky pools in association of *Dictyota dichotoma* (Hudson) Lamx., *Dilophus fasciola* (Roth) Howe, *Taonia atomaria* (Woodw.) J. Ag. and *Laurencia*
spp. This new species also grows in association of *Posidonia
oceanica* (L.) Delile. In this association it is much more protected resulting in bearing broader leafy segments than those

of other associations. Along the Libyan coast it has been found
growing during the months of March to September but according
to the local fishermen the species grows throughout the year in
deep water.

The specimens from El-Khoms, 100 km east of Tripoli differ from
those of other localities in possessing 3 layers of cells
throughout the wing of the mature segments. This needs further
investigations.

D. tripolitana resembles *D. australis* f. *karachiensis* Nizam. &
Saif. in habit, in possessing transversely obliquely arranged
sori or crypts on either side of the midrib but differs in
thinly arranged sori as well as in thin texture.

Dictyopteris membranacea (Stackhouse) Batters, described and
figured by Segawa (1960) Pl. 15, fig. 129 also appears to be
similar to *Dictyopteris tripolitana* as the sporangia are transversely obliquely arranged along the midrib.

DICTYOPTERIS POLYPODIOIDES Lamouroux
Plate III B, IV, V, XXIV. Figs a, b, c, D, E, F, G.

Synonyms: *Fucus membranaceus* Stackhouse 1795: 13 (excl. Pl. VI). *Fucus membranaceus* Turner 1809: 41, pl. 87. *Haliseris polypodioides* (Desfont.) J.
Agardh of Harvey 1846-51: pl. 19, figs. 1, 4; 1849: 36, pl. 6 B. Gray 1867:
57, pl. 3, fig. 3. Crouan et Crouan 1867: 169, pl. 29. *Haliseris polypodioides* (Lamx.) C. Agardh 1820: 142; 1821: 142; 1823: 142; 1824: 262. J.
Agardh 1842: 36. Ardissone e Strafforello 1877: 165. De Toni et Forti 1914:
291. De Toni et Levi 1888: 248. Greville 1830: 64. Grunow (in Heufler)
1861: 425. Hooker 1844: 286. Kützing 1849: 561. Langenbach 1873: 14. Meneghini 1842: 252. Montagne 1840: 145. Muschler 1910: 302. Naccari 1828: 78.
Pampanini 1916: 290; 1931: 35. Piccone 1892: 48. Schiffner 1926: 3o6 (incl.
forms). Zanardini 1847: 50. *Dictyopteris membranacea* (Stackh.) Batters of
Hamel 1939: 339, fig. 56i. *Dictyopteris polypodioides* Lamx. var. β *angusti-
frons* Mont. 1846: 30. *Neurocarpus membranaceus* Weber et Mohr 1805: 300.

Bory 1832: 321; 1838: 75. Debray 1893: 10; 1897: 48; 1899: 97. Duby 1830:
954. Gaillon 1828: 370. Lamouroux 1809b: 131; 1813: 271. Montagne 1846: 28.
Ollivier 1929: 117.

Holdfast discoid with velvety fibrous rhizoids. Fronds erect, thin, flexuous, to 22 cm high, laterally or sub-dichotomously branched into leafy segments. Segments linear, thin, membranous, flat, to 8 mm broad, to 14 cm long, 850 µm thick, percurrent midrib 1-2 mm broad, reaching up to apices; composed of many layers of cells in thickness. Wings strongly wavy; margin entire but sometimes strongly lacerated, composed of 2 layers of central cells, 60-80 µm thick (mature segments). Basal parts 2-5 mm broad, to 500 µm thick and denuded of wings. Proliferations very common from or near the midribs. Young segments linear, oblong to 3 mm broad; midrib to 1 mm broad and to 450 µm thick; margin entire; apices acute; composed of 2 layers of cells to 50 µm thick. Sori or crypts in linear rows along the midrib on either side. The linear arrangement of sori very prominent on proliferated leafy segments (proliferations). Occasionally oblong patches of sori occur along the midrib.

Specimens examined: TUNISIA - Carthage (Leg. B. Schussnig, 29. 4.1924, Berol. No. 24690, as f. *angustior* Schiffn.). Between La Marsa and Sidi-Bou-Said (Leg. B. Schussnig, 17.4.1924, Berol No. 24693 as f. *angustior, e costa prolifera* Schiffn.). Gulf of Gabez (Leg. J. Gerloff, 9.8.1969, Berol. No. 25692).
LIBYA - Tripoli, near Ender Hospital (Leg. M. Nizamuddin, 15.6 1974, No. 0512; 9.9.1974). Tripoli, Sciare Sciat (Leg. M. Nizamuddin, 11.11.1974, No. 1245).
YUGOSLAVIA - Zara (Herb. Stackmayer ex herb. Grunow, Berol. No. 10721). Rovinj (Leg. Krumbach, 28.11.1914, Berol. No. 10706). Yugoslavia (Leg. Lokrum, 14.9.1954, Berol. No.10704).
ITALIY - Trieste (Leg. J. Schiller, 20-25 September 1924, Berol No. 24691 as *Haliseris polypodioides* (Desfont.) J. Ag. f. *angusta* Schiffn. Leg. Cori, 20.11.1914, Berol. No. 08612).
SICILY - Siracusa (Leg. Jauchen, 6.4.1913, Berol. No. 10708). Trapani (Leg. G. Langenbach, 11.9.1871, Berol. No. 10725 A).
FRANCE - Rade de Brest (Leg. J. Gerloff, 15.9.1962, Berol. No. 17646). Cannes (Hamel's specimen no. 25, 1927, in BM). Gŭéthary (Leg. J. Gerloff, 22.9.1962, Berol. No. 04859). Antibes (Leg. L. Kny, December 1863, Berol. No. 24626).
MALLORCA - Magaluf (Leg. J. Gerloff, 17.2.1970, Berol. No. 25824). Palma Nova (Leg. J. Gerloff, 17.2.1970, Berol. No. 25808).

PORTUGAL - (cf. Rodrigues' plate no. Va, 1963).

ENGLAND - Plymouth (Leg. T. Gatcombe, Berol. No. 10731. Leg. J. L. William ex herb. Holmes, 8.9.1905, in BM). Sidmouth (Leg. Holmes, 6.9.1892, in BM). Brook (Leg. Holmes, 8.9.1896, in BM). CANARY ISLANDS - Punta del Hidalga, Teneriffe (Leg. J. Gerloff, 21.10.1973, Berol. No. 29101). U.S.A. Bogue Banks, North Carolina (Leg. J. Kohlmeyer, 26.7.1963, Berol. No. 23380).

Dictyopteris polypodioides is most probably a deep water alga along the Libyan coast as it has always been found as drift. *D. polypodioides* resembles *D. membranacea* in general habit but differs in possessing flexuous fronds with linear arrangement of sori along either side of the midrib.

PADINA Adanson 1763

Fronds tufted erect or procumbent attached by compact felted rhizoidal holdfast; stipitate often invested by rhizoids; upper part fan-shaped, splitting into lobes, apices curved, involuted; composed of two to several layers of cells in thickness; zonate, zones marked by concentric rows of hairs; calcification on one or both surfaces; sporangia singly or scattered or in sori between or above the hair lines, indusiate or non-indusiate; antheridia and oogonia also in sori between the hair lines — oogonial sori indusiate, antheridial ones sometimes non-indusiate. Sporangia producing 4 spores. Stipe occasionally develop coarse, ramified prostrate or procumbent laterals (termed growth phase) branches of which ultimately become flat and fan-shaped. Growth phase laterals grow vegetatively by means of a single apical cells and the flattened parts by marginal initials.

Type: *Padina pavonia* (L.) Lamouroux 1816: 304. (*Fucus pavonius* L.; [*Fucus pavonicus* L.; *Padina pavonica* (L.) Thivy].

This genus is easily recognizable but the distinction between the species has been the source of much confusion in the past and still holds true in some cases. The confusion or the difficulties are due to the morphological similarities among the species of this genus. Hauck (1887) discussed the difficulties

23

in the identification of the species of *Padina* from Indian
Ocean and the Red Sea. Papenfuss (1968) also realised this dif-
ficulty and suggested that there may be more species involved
in *Padina pavonia* complex. Presently the author met with the
same difficulties. In the present work five characters were
used which were useful for the separation of the species of
this genus - (i) the mutual arrangement of the sori and the
lines of hairs, (ii) the presence or absence of an indusium and
(iii) the number of cell layers in the thallus, (iv) hairs or
sori on both surfaces or only on the upper surface or on the
lower surface, (v) habit of the fronds. The flat and the
flabellate parts of fronds grow vegetatively by means of marginal initials which is a complex of apical cells of a complex
filaments. Growth phase arising from the stipe possesses an
apical cell (Plate XXVI, Fig. 16). The apical cell persists
till the growth phase remains terete but disappears as it be-
comes flat or compressed. Apical growth is taken over gradually by marginal initials. The growth phase composed of many
layers of large central cells but decreased to one layer of
large central cells as the upper parts become compressed or
flat. Throughout the world so far 31 species of *Padina* has
been reported. In the past a single species, *Padina pavonia*,
was reported from the Mediterranean Sea. A thorough survey
along the coast of Libya reveals the occurrence of two additional species of *Padina*, *P. gymnospora* (Kg.) Vick. and *P. tenuis* Bory.

KEY TO THE SPECIES

1. Indusium present . . . . *P. pavonia*
1. Indusium lacking 2

2. Sori between the hair lines . . . . *P. tenuis*
2. Sori in the middle of each alternate zone of hair lines . *P. gymnospora*

24

PADINA PAVONIA (L.) Lamouroux

Plate VI, XXVIII. Figs. 18-23, a.

Basionym: *Fucus pavonius* Linnaeus 1759: 1345; 1763: 1630; 1764: 1630 (= *Fucus pavonicus* Linnaeus 1753: 1162.

Synonyms: *Ulva pavonia* Linnaeus 1767: 719; 1770: 719; 1774: 817. Desfontaines 1798: 428. Esper 1797: Ulv. t. 4. Hooker 1844: 285. Hudson 1798: 566. Lamarck et de Candolle 1805: 17. Roth 1800: 240. Sowerby and Smith 1797: 1276. *Dictyota pavonia* Lamouroux 1809a: 39; 1809c: 331; 1813: 272. *Zonaria pavonia* C. Agardh 1817: XX; 1820: 125; 1821: 125; 1823: 125; 1824: 263. Kützing 1845: 272; 1849: 565; 1859: 28 t. 70. Trevisan 1849: 463. *Padina mediterranea* Bory 1827: 590; 1832: 320; 1838: 75. *Padina oceanica* Bory 1827: 590.

J. Agardh 1842: 39; 1848: 113; 1882: 119. Ardissone 1883: 486. Ardissone e Straforello 1877: 165. Ardrê 1969: 399. Batters 1902: 54. Børgesen 1926: 86. Bornet 1892: 230. Bory 1828: 145. Crouan et Crouan 1867: 169. Debray 1893: 10; 1897: 48; 1899: 96. De Toni et Levi 1888: 478. Duby 1830: 955. Feldmann 1931: 217; 1937: 317. Funk 1927: 365; 1955: 51. Gaillon 1828: 371. Gayral 1958: 230; 1966: 261. Gerloff und Geissler 1971: 752. Giaccone 1968: 223; 1969: 500. Gray 1867: 57. Hamel 1939: 343. Harvey 1846-51: Pl. XCL; 1849: 47; 1852: 104. Harvey-Gibson and Knight 1913: 306. Hauck 1885: 309; 1887: 42. Lamouroux 1816: 304. Langenbach 1873: 14. Levring 1974: 32. Le Jolis 1863: 98. Mazza 1903: 4; 1904: 120. Meneghini 1842: 239. Montagne 1840: 145; 1846: 33. Nasr 1947: 77. Nizamuddin and Lehnberg 1870: 122. Ollivier 1929: 17. Pampanini 1931: 35. Reinbold 1898: 89. Reinke 1878: 15. Rodrigues 1963: 44. Schiffner 1926: 306. Srinivasan 1973: 21. Taylor 1960: 234. Zanardini 1847: 5o. Zinova 1967: 145.

Holdfast fibrous and discoid giving rise to erect fronds to 7 cm high, stipate, stupose and calcified. Stipe erect to 1 cm high, terete to compressed to 1.5 mm broad, 370-400 µm thick, composed of 6 layers (± 1) layers of cells in thickness - 4 layers of large central cells and covered on each side by a single layer of peripheral cells. Numerous rhizoids develop from the peripheral cells of the stipe. Fronds flat, flabellate to 24 cm broadly expanded, splitting into lobes, zonate at a distance of 2 mm; cuneate below; composed of 3 layers of radially elongated cells, 150-200 µm thick, Apical and marginal

25

parts composed of 2 layers of flat cells to 50 μm thick. Upper
surface strongly calcified, lower surface slightly calcified.
Calcification 50-70 μm thick. Fertile bands 1-2 mm broad. Sterile ones 3-5 mm broad. Sori on the ventral surface of the frond
on either side of each hairlines close to it, with persistent
indusium, producing 4 spores, ovoid 100-130 x 70-100 μm. Monoe-
cious.

Specimens examined: MOROCO - Tanger (cf. Debray, 1897).
ALGERIA - Tipasa, Cape Caxine; St. Eugène; Mole sud Alger;
Tenés; Cherchell; Oran; Stora; Philipeville; Bone (cf. Debray,
1897).
LIBYA - Jodain (Leg. M. Nizamuddin, 4.5.1973). Andulus Beach
(Leg. Gadeh, 10.7.1973, No. 0545). Gargaresh (Leg. M. Nizamud-
din, 13.5.1974, No. 0422, 2 m depth. Tripoli near Govt. Guest
House (Leg. M. Nizamuddin, 12.5.1971, No. 0086). Tripoli, off
Ender Hospital (Leg. M. Nizamuddin, 10.4.1972, No. 0109; 15.4.
1974). Ghote El Roman (Leg. M. Nizamuddin, 22.6.1974, No. 0544)
Tripoli and Syrte (Leg. Petersen, 26.7.1910, Station No. 150).
Tripoli (Leg. Muschler, 1910). Bengazi (Leg. Hainmann, 1881.
Ruhmer, 1882). Bengazi-Guiliana (Leg. Zanon, 1915).

Padina pavonia grows in very varied habitats ranging from upper
to sub-littoral regions, even in polluted conditions. In calm
condition the fronds were infested with crustose algae but in
exposed condition it was inhabited by *Enteromorpha* spp. and
Cladophora spp. Kützing (1859) figured two forms, *Zonaria pavonia anglica* Kg. and *Z. pavonia neapolitana* Kg. These two
forms resemble each other anatomically in possessing 3 layers
of cells in the upper parts but differ in lower parts. These
are two distinct forms of *Padina pavonia*. The new combinations
stand as *Padina pavonia* f. *anglica* (Kg.) comb. nov. (Kützing,
1859, Tab. Phycol. IX, 28, t. 70, f. I) and *P. pavonia* f. *neapolitana* (Kg.) comb. nov. (= Kützing, 1859, Tab. Phycol. 28,
t. 70, f. II). Libyan specimens resemble *P. pavonia* f. *anglica*

PADINA GYMNOSPORA (Kützing) Vickers

Plate VIII, XXV. Figs. 1-5.

Basionym: *Zonaria gymnospora* Kützing 1859: 29, tab. 71. Fig. IIa, b, c.

Synonym: *Padina australis* Hauck 1887: 44. Weber van Bosse 1913: 180; 1928: 490.

Børgesen 1914: 202; 1930: 170; 1937b: 315; 1941: 49; 1948: 47. Chapman 1963: 32. Hoyt 1920: 458. Taylor 1960: 237. Thivy 1959: 69. Vickers 1905: 58; 1908: 37, Pl. 7. Misra 1966: 156.

Holdfast discoid to 7 mm across, with fibrous rhizoids. Stipe erect to 5 mm long, compressed to 3 mm broad. Fronds erect to 16 cm high, cuneate below, and upper part 18 cm in expanse, flat, flabellate, membranous, concentrically zonate; upper surface more calcified than the lower one; zonation at a distance of (3-)4 mm. Fronds splitting into a number of lobes, 100-200 μm thick. Terminal part composed of 3 layers of cells, 100-120 μm thick. Middle part composed of 4 layers of cells, 150-200 μm thick. Stipe or lower parts composed of 3 layers of cells, 160-200 μm thick. Every row of tetrasporangia with a row of hairs on either side leaving a broader sterile part between the rows of hairs. Tetrasporangia ovoid, 70-110 x 50-60 μm. Spores round to 100 μm diam.

Specimens examined: Sciare El Fateh, Tripoli (Leg. Dr. M. Nizamuddin, 13.6.1974, No. 0517).

This species generally grows in upper littoral rocky pools in association of *Cystoseira compressa* (Esp.) Gerloff & Nizam. and *Cystoseira discors* (L.) C. Ag. emend. Sauv. The species was also collected from 5 m depth.

Libyan specimens resemble the type in habit and description, but differ in possessing 4 layers of cells in the middle part of the fronds whereas the type possesses 4 layers of cells in the lower parts.

27

PADINA TENUIS Bory

Plate VII, XXV-XXVII. Figs. 81, 82, 6-16, 1-6.

Synonyms: *Zonaria pavonia* δ *tenuis* C. Agardh 1824: 264. *Zonaria tenuis* β *commersonii* Kützing 1849: 565. Montagne 1863: 67.

J. Agardh 1848: 113; 1882: 119. Børgesen 1941: 49; 1948: 48; 1953: 13; 1930: 170; 1936: 77; 1937b: 315; 1939: 32. Bory 1827: 591; 1828: 144, Pl. 21, Fig. 2. Weber van Bosse 1913: 178; 1928: 490. Hauck 1887: 44. Misra 1966: 155. Setchell 1926: 92. Srinivasan 1973: 22.

Fronds, attached by discoid holdfast with fibrous rhizoids, stupose, stipate, Stipe erect to 1 cm long, 1-2 mm broad and 0.5-1 mm thick. Fronds generally in tufts, erect to 10 cm high, cuneate below; flat, flabellate, membranous, zonate above; upper surface strongly calcified (100 μm thick); apices in-rolled with projecting hairs; terminal parts splitting into number of lobes. Near the apices the frond is composed of 2 layers of cells to 50 μm thick; downwards composed of 3 layers of cells, (100-)120-150(-160) μm thick. Stipe composed of 3 layers of cells, 100-130 μm thick, surrounded by dense branch-ed rhizoids. The transition zone, the region between the stipe and the lower part of the lobes composed of 3 or 4 layers of cells - one large central cell alternating with two layers of central cells. The transition region is composed of 1 or 2 layers of large central cells. In longitudinal sections the peripheral cells of the upper surface are larger than those of the lower one. Central cells are uniform, 70 μm broad and 40 μm thick (high). Every other rows of hairs well developed. Fertile and sterile zones alternate each other. Fertile zone 0.2-1 mm broad. Sterile zone 2-3 mm broad. Tetrasporangia stalked, 70-170 x 50-110 μm. Spores round, 90-120 μm. No in-dusium present.

Specimens examined: LIBYA - Sciare El-Fateh, Tripoli (Leg. M. Nizamuddin, 13.6.1974). Gargaresh (Leg. M. Nizamuddin, 13.5. 1974, No. 0496, 2 m depth). El Khoms (Leg. M. Nizamuddin, 12. 4.1974, No. 0358) flat rocky pools on exposed side of boulders No. 0497, upper littoral pools).
CRETE - Chersonissos (Leg. J. Gerloff, 29.5.1974, Berol. No.

29356, below 1 m depth on rocks).

GREECE - Kyllini, Peloponios (Leg. J. Gerloff, 15.5.1975, Berol. No. 29713, in 2 m depth on rocks). Corfu, Paleokastritza (Leg. Kuhbier, May 1968, in 1 m depth, Berol. No. 24660). ITALY - (Leg. J. Schiller, 7.9.1911, Berol. Nos 10798, 25033). Adriatic (ex herb. Runs., Berol. No. 10808). Cigale, Lessina Island (Leg. Engelhardt, 4.6.1935, Berol. Nos. 10810, 10787). St. Nazzaro, near Genova (Leg. Ardissone, July 1857, Berol. No. 10815). Livorno (ex herb. Bauschianum, Berol. No. 10799). SICILY - Palermo (Leg. L. Kny, 2.3.1870, Berol. Nos. 10818, 10786A; 27.4.1870, Berol. No. 10820). Arenella (Leg. Langen-bach, 16.6.1870, Berol. No. 10819; 23.5.1870, Berol. No. 10786 B, C, D; 16.6.1870, Berol. No. 10822; 16.6.1870, Berol. Nos. 10816, 10790). Aspra (Leg. G. Langenbach, 8.6.1870, Berol. No. 10785).

YUGOSLAVIA - Quarnero, Lovrana (Leg. . . . June 1882, Berol. No. 25034). Abbazia, Istria (Leg. J. Brunthaler, July 1905, Berol. Nos. 10794, 30659, ex herb. W. Migula). Ikam, Istria (Leg Zay, Berol. No. 10783). Rovinj (Leg. J. Gerloff, 7.6.1961, Berol. No. 20052, 8-1.5 m depth; 15.6.1961, Berol. No. 19904, in polluted water; 15.6.1961, in polluted water, Berol. Nos. 19901, 19902, 19903. Leg. A. Vatova, 8- 1932, Berol. No. 08610.

FRANCE/CORSICA - Bastia beach (Leg. Fr. E. Ronniger, 4.6.1914, Berol. No. 25045, as *P. pavonia* var. *tenuis* Kg.).

GREECE - Syra (Leg. Liebetruth, 1870, Berol. No. 10779). CRETE - North coast near Candia (Leg. B. Watz, 24.4.1914, as *P. pavonia* f. *tenuis* Kg.).

YUGOSLAVIA - Dubrovnik, Lokrum Is. (Leg. Pittirsch, Yugoslavian excursion, 15.9.1954, Berol. No. 10784, 1-2 m depth). Valesca near Veglia, Veglia Is. (Leg M. Lusina, 29.4.1915, Berol. No. 10806 as *P. pavonia* f. *nana* Schiff.). Ragusa (Leg. V. Schiff-ner, 5.4.1926, Berol. No. 25028, as *P. pavonia* f. *nana* Schiff.). Val di Bora, Rovinj (Leg. V. Schiffner, 4/8 July 1914, Berol. No. 25024).

TUNISIA - Soliman, Bay of Tunis (Leg. J. Gerloff, 19.9.1967, Berol. Nos. 24180, 24181). Gulf of Gabes (Leg. J. Gerloff, 11. 3.1969, Berol. No. 25699).

GREECE - Katapola, Amorgos Is. (Leg. K.H. Rechinger, 6 & 7, 932, Berol. No. 10801). Kephalonia Is. (Leg. W. Schimper,

1804, Berol. No. 10781).

GREECE - Attika, rocky coast of Cape Sounion (Leg. J. Gerloff, 7.4.1971, Berol. No. 26983). West side of Cape Sounion (Leg. J. Gerloff, 24.5.1973, Berol. No. 28835).

GRETE - Chersonissos (Leg. J. Gerloff, 29.5.1974, Berol. No. 29349; 28.5.1974, Berol. No. 29347). Vai (Leg. J. Gerloff, 7.6. 1974, Berol. No. 29375). Stalis (Leg. J. Gerloff, 30.5.1974, Berol. No. 29390; 11.6.1974, 29400). North coast near Candia (Leg. B. Watzl., 24.4.1914, Berol. No. 25025, as *P. pavonia* f. *neapolitana* Kg.).

YUGOSLAVIA - Pelagosa Is. (Leg. J. Schiller, July 1912, Berol. No. 25036). Lissa, Lissa Is. (Leg. V. Schiffner, 17.4.1928, Berol. No. 10805). Ikam, Istria (Leg. Zay, ex herb. Flora exsiccata Austro-Hungarica, Berol. No. 10783). Split (Leg. . . . Aug. 1900, Berol. No. 10809; Leg. Titius, Berol. No. 10821). Rovinj (Leg. V. Schiffner, 28.11.1914, Berol. No. 10804). Val di Bora, Rovinj (Leg. J. Brunnthaler, 24.5.1913, Berol. No. 10826). Figarola, Rovinj (Leg. J. Gerloff, May 1970, Berol. No. 26301).

ITALY - Adria (ex herb. Stockmayer (ex herb. Grunow), Berol. No. 25038), Berol. No. 25038). Barcola (Leg. F. Krasser, Kryptogamae exsiccatae, Berol. No. 10793). Portofino, east Liguria (Leg. F. Baglietto, ex Erb. Critt. Itall. ser. II, Berol. No. 10797). San Montane (Leg. . . . . May 1895, Berol. No. 27977). St. Andrea (Leg. Engelhardt, 19.1.1882, Berol. No. 10811). Lymonn, Lessina (Leg. . . . . Aug. 1844, Berol. No. 25035). SICILY - Trapani (Leg. G. Langenbach, 5.9.1871, Berol. No. 10817; Berol. No. 10791A). Arenella (Leg. . . . . Berol. No. 10791B).

FRANCE - Caines, St. Marguerite Is., Ligur Sea (Leg. G. Ronniger, June 1925, Berol. No. 25030). St. Malo, Cape du Troc (Leg. Runze, 11.6.1962, Berol. No. 19874). Banyuls (Leg. J. Kohlmeyer, 23.5.1962, Berol. No. 17594; Leg. Runze, 6.9.1962, Berol. No. 17921). Antibes (Leg. T. Broeksmit, April 1925, Berol. No. 25025. Leg. L. Kny, December 1860, Berol. No. 10791C). Guéthary (Leg. Gerloff, 22.9.1961, Berol. No. 04921). Hendaye (Leg. J. Gerloff, 24.9.1961, Berol. No. 04965).

SPAIN/MALLORCA - Paguera (Leg. J. Gerloff, 20.2.1970, Berol. No. 25800). Megaluf (Leg. J. Gerloff, 17.2.1970, Berol. No. 25823).

TUNISIA - Tabarka (Leg. J. Gerloff, 11.4.1968, Berol. No. 24516). Ain Oktor (Leg. J. Gerloff, 10.4.1968, Berol. No. 24515).

The species grows in low lying rocky pools in littoral zones along very exposed shore. In some regions the species grows in very strong rough condition subject to strong wave action. In exposed condition the fronds become strongly calcified and free from epiphytes. Stipes occasionally develop an extensive coarse, fibrous branched growth phase. The branches are tereto-compressed, prostrate, 450-600 μm broad and 200-250(-320) μm thick. The branches of the growth phase are variously branched either dichotomously or laterally or irregularly and gradually becoming flat, flabellate, 3 mm broad to 250 μm thick and com-posed of 4 layers of cells to 150 μm thick; middle parts com-posed of 3 layers of cells and apices composed of 2 layers of cells. Terete branches grow vegetatively by means of an apical cell but as soon as these become flat, flabellate apical growth is replaced by marginal initials. The growth phase is very com-mon in this species.

Fronds erect, tufted, arising from an expanded felted rhizoidal
holdfast; cuneate below, flat above; branches plane, cleft into
lobes; lobes transversely banded by rows of hairs. Stipe con-
tinued to some distance into lobes as midribs or as thickened
central part; vegetative growth by means of marginal intials;
lobes or fronds composed of many layers of regularly centrally
placed cells, each of which is covered by a single layer of
peripheral cells on each surface. Sori zonately arranged containing sporangia, producing 8 spores, and paraphyses.

Type: *Zonaria flava* (Clemente) C. Agardh. (*Fucus flavus*
Clemente syn. tax. *Z. tournefortii* (Lamouroux) Montagne [*Fucus
tournefortii* Lamouroux].

Zonaria flava (Clemente) C. Ag. is the only species found along
the coast of Libya as well as in other parts of the Mediterranean Sea. C. Agardh (1817, 1820, 1821, 1823, 1824) always
accredited the genus to Draparnaud but he was not justified in
doing so. He himself is to be accredited not others. *Zonaria* C.
Ag. 1817 has already been conserved by the International Code
of Botanical Nomenclature (Lanjouw, 1952; Stafleu, 1972, 1978)
though Silva (1952) vehemently criticized in accrediting C.
Agardh instead of J.G. Agardh (1841).

ZONARIA FLAVA (Clemente) C. Agardh
                Plate IX, XA; XXIX, XXXI. Figs. 76-80, tab. 336, 4.

Basionym: *Fucus flavus* Clemente 1804: ? ;   1807: 310, non Linnaeus 1781:
452; 1792: 1390.

Synonyms: *Fucus tournefortii* Lamouroux 1805: 44, t. 26, fig. 1. *Dictyota
tournefortiana* Lamouroux 1809a: 40; 1809c: 331; 1813: 272. *Padina tournefortiana* Duby 1830: 955. Montagne 1840: 146. *Phycopteris tournefortii*
(Lamx.) Kützing 1859: 26, t. 65, fig. 1. *Zonaria tournefortii* (Lamx.) Montagne 1846: 32. Børgesen 1926: 92; 1938b: 227. Bornet 1892: 236. Debray
1893: 10; 1897: 48. Gayral 1958: 225. Gerloff und Geissler 1971: 753.
Giaccone 1969: 501. Hamel 1939: 338. Levring 1974: 33. Simons 1964: 195.

Phycopteris dentata Kützing 1859: 26, t. 65, Fig. II. *Orthosorus tournefor-tii* Trevisan 1849: 459. *Orthosorus stuposus* Trevisan 1849: 459.

C. Agardh 1817: XX; 1820: 130; 1821: 130; 1823: 130; 1824: 265. J. Agardh 1841: 444; 1848: 119; 1894: 13. Ardissone 1883: 490. De Toni 1895: 230. Funk 1927: 366; 1955: 52. Giaccone 1968: 223. Harvey 1858: 123. Hoyt 1920: 455. Mazza 1904: 119. Meneghini 1842: 235.

Fronds erect 5-22 cm high, cuneate below, flat and flabellate above, splitting into lobes. Stipe short 2-3 cm high, clyin-drico-terete, 3-5 mm across. Lobes alternately arranged, to 13 cm long, flat, flabellate to 6 mm broad ecostate; lower parts costate or centrally strongly thickened; zonate above at a distance of 2 mm (in some cases faintly zonate). Apices com-posed of 4 layers of cells in rows increasing downwards to 6 layers of cells. Peripheral cells of same width and single layered on either side of the surface containing dense chroma-tophores. Fronds 80-100 μm thick i.e. central layer (60-)70-80 μm thick and each peripheral layer 10-15(-20) μm thick or 1/2 the width of the medullary cells. Sori indusiate generally in zones but may be irregularly distantly distributed or scat-tered containing sporangia, on either surface. Sporangia pro-ducing 8 spores without stalk. Paraphyses occur in sori.

Specimens examined: MOROCCO - Tanger (Leg. M. Askenasy. cf. Bornet 1892).
ALGERIA - St. Eugene (Leg. F. Debray, June 1887, Berol. No. 10866).
TUNISIA - Between Gulf of Gammarth and mouth of Med Jerda, sta. 1 (Leg. Cherif, 11.12.1973, muddy bottom, 43-5 meter).
LIBYA - Tripoli (cf. Muschler 1910). Bengazi (Leg. M. Nizamud-lin, 3.3.1973, No. 0544. Drift).
GREECE - Paleokastritsa, Corfu (Leg. Kuhbier, May 1968, Berol. No. 24657).
ITALY - Gryosta di Misenum, Gulf of Naples (Leg. J. Brunnthaler, 29.4.1914, Berol. No. 10868B).
SICILY - Catania (cf. Mazza, 1904).
FRANCE - Antibes (Leg. L. Kny, December 1863, Berol. No. 10870).
CANARY ISLANDS - Gran Canaria (Leg. Th. Liebetruth, 1863, Berol. No. 10865 as *Phycopteris stuposa* Kg.: herb. Stockmayer ex herb.

Grunow 1863, Berol. No. 10867 as *Zonaria tournefortii* f. *flava*
J. Ag.). Punta del Hidalgo, Teneriffe (Leg. J. Gerloff, 21.10.
1973, Berol. Nos. 29089, 29102; on rocks with *Cystoseira abies-
marina* (Turn.) C. Ag.). Between Punta del Hidalgo and Bajamar,
Teneriffe (Leg. J. Gerloff, 7.11.1973, Berol. No. 29141). El
Medano, Teneriffe (Leg. J. Gerloff, 3.10.1973, Berol. No.
29131). Puerto de la Cruz, Teneriffe (Leg. J. Kohlmeyer, 29.11.
1964, Berol. Nos. 23921, 23922, 23923).
MADEIRA - (Leg. M. Rodrigues, March 1965, Berol. No. 10869).

Zonaria flava is a very rare species found along the Libyan
coast. The species has been collected twice. It appears to be
a deep water alga as also reported by Funk (1927) from Naples
and Taylor (1960) reported it from a depth of 50-90 m from the
eastern American coast. It was collected from Tripoli and
Bengazi during the month of March along the European coast of
the Mediterranean Sea through May to October. Libyan specimens
differ in structure from those of the Canary Islands (Børgesen,
1926) but resemble those of the European coast of the Mediterranean Sea (Gayral, 1958. Hamel, 1939. Kützing, 1859).

Kützing (1859, Tab. Phycol. 9: 26, t. 65) reported two varieties, *Phycopteris tournefortii* (Lamx.) Kg. var. *latifolia* Kg.
and *Phycopteris tournefortii* var. *palmatifida* Kg. Lack of suf-
ficient material do not warrant any comment from the author.
[Now these varieties stand as *Zonaria flava* var. *latifolia* (Kg.
comb nov. and *Z. flava* var. *palmatifida* (Kg.) and comb. nov.].

TAONIA J. Agardh 1848

Holdfast discoid having tufted rhizoids. Fronds erect, cuneate
below, flat above, ecostate, irregularly lobed or segmented,
apices truncated or fimbriated; lobes or segments thick or thin
in texture, zonate. Sori, containing tetrasporangia, generally
zonate or occasionally scattered on both surfaces; hairs also
in zones; gametangia in zones or scattered; plants dioecious.

Type: *Taonia atomaria* (Woodward) J. Agardh. (*Ulva atomaria*
Woodward).

Taonia atomaria is the only species belonging to this genus found along the entire coast of the Mediterranean Sea. Two distinct forms occur along the coast.

KEY TO THE FORMS

Lobes or segments with entire margin . . . . *T. atomaria* f. *atomaria*
Lobes or segments with strongly dentate margin . . . *T. atomaria* f. *ciliata*

TAONIA ATOMARIA (Woodward) J. Agardh f. *ATOMARIA* J. Agardh
Plate X B, XI A, XIII, XXX, XXXI. Figs. 29-31, 2.

Basionym: *Ulva atomaria* Woodward 1797: 53.

Synonyms: *Fucus zonalis* Lamouroux 1805: 38, t. 25, fig. 1. *Zonaria atomaria* (Woodward) C. Agardh 1820: 128; 1821: 128; 1823: 128; 1824: 264. Gray 1821: 341. Greville 1824: 298. *Padina phasiana* Bory 1832: 320; 1838: 75. *Padina atomaria* Montagne 1840: 146. *Stypopodium flavum* Kützing 1843: 341; 1849: 563; 1859: 25, t. 62, fig. II. *Stypopodium attenuatum* Kützing 1849: 563; 1859: 25, t. 63, fig. II. *Stypopodium atomaria* Kützing 1843: 341; 1849: 563; 1859: 24, t. 61, fig. a. Ardissone et Straforello 1877: 164. *Dictyota zonata* Lamouroux 1809a: 40; 1809c: 331; 1813: 272. *Dictyota atomaria* Greville 1830: 58, non Hauck 1887. J. Agardh 1842: 37. Hooker 1844: 284. Meneghini 1842: 229. Reinke 1878: 26. Zanardini 1847: 50. *Dictyota atomaria* var. *nuda* Schiffner 1931: 189.

J. Agardh 1848: 101; 1894: 29. Ardissone 1883: 483. Ardré 1969: 398. Batters 1902: 54. Børgesen 1926: 89. Bornet 1892: 229. Crouan et Crouan 1867: 168. Debray 1893: 10; 1897: 48; 1899: 96. Feldmann 1937: 318. Funk 1955: 50. Gayral 1958: 232; 1966: 263. Gerloff und Geissler 1971: 753. Giaccone 1969: 501. Gray 1867: 59. Hamel 1939: 337. Harvey 1849: 38. Hauck 1885: 307. Langenbach 1873: 14. Le Jolis 1863: 98. Levring 1974: 32. Mazza 1903: 4; 1904: 120. Muschler 1910: 302. Newton 1931: 213. Ollivier 1929: 117. Pampanini 1931: 34. Piccone 1892: 48. Rodrigues 1963: 57. Robinson 1932: 113.

Holdfast discoid with fibrous rhizoids. Fronds erect to 17 (-30) cm high, cuneate below, flat above, to 1(-2) cm broad, membranous, to 150 μm thick, dichotomously or irregularly

35

branched or upper parts strongly splitted into lobes, concentrically zonate, margin entire (rarely distantly dentate) composed of many layers of cells. Central layers of broad, polygonal cells covered on either side by a single layer of broad, radially arranged peripheral cells; cell walls thickened. Sori in prominent zones or scattered below on both surfaces. Zones sometimes wavy. Sori containing tetrasporangia producing 4 spores. Indusium lacking.

Specimens examined: MOROCCO - Tanger (. . . 26.7.1879 as *Stypopodium attenuatum* in K).
LIBYA - Tripoli, near Jamarak (Leg. M. Nizamuddin, 23.6.1973).
Tripoli, near Govt. Guest House (Leg. M. Nizamuddin, 12.5.1971)
El-Khoms (Leg. M. Nizamuddin, 12.4.1974; 23.5.1974; 15.6.1973; 23.6.1973).
CRETE - Stalis Leg. J. Gerloff, 30.5.1974, Berol. Nos. 20042, 29391; 11.6.1974, Berol. No. 29396). Chersonisos (Leg. J. Gerloff, 28.5.1974, Berol. No. 29384).
YUGOSLAVIA - Bocche di Cattaro (Leg. J. Schiller, 30.5.1912, Berol. No. 25061). Near Orebic, Sabbioncello Peninsula (Leg. K. H. Rechinger, April 1939, Berol. No. 10855). Rovinj (Leg. F. Baron de Lichtenstein, Krypt. Exs. No. 1510, Berol. No. 10849, from 2 m depth; Berol. No. 25062. Leg. J. Gerloff, May 1970, Berol. Nos. 26299, 26366; 15.6.1961, Berol. No. 24670).
ITALY - Cape Misenum, Gulf of Naples (Leg. J. Brunnthaler, 29. 4.1914, Berol. Nos. 10854, 08609 as *Taonia atomaria* f. *nana* Schiff. nom. nud.) Naples (Leg. Pedioiro in K). Posilipo, Naples (Leg. Mazza, 1904). Ischia (Leg. J. Corinaldi, Berol. No. 10850). Porte Savona (Leg. A. Piccone, April 1860; April 1862 in K).
SICILY - Palermo (Leg. L. Kny, 23.3.1870, Berol. No. 10845 B; 27.4.1870, Berol. No. 10857; 7.4.1871, Berol. No. 10856). Fra Torre, Acireal (Leg. Mazza, 1904).
FRANCE - Nice (Leg. . . . 26.6.1879 ex herb Wollny, as *Stypopodium flavum,* in K). Banyuls, towards Cerbère (Leg. J. Kohlmeyer, 5.6.1962, Berol. No. 17583. Leg. Runze, 8.9.1962, Berol. No. 17958). St. Vaast (Leg. . . . 2.10.1853, ex herb. G. Thuret in K). Rade de Brest (Leg. Runze, 15.9.1962, Berol. No. 17641) Hendaye (Leg. J. Gerloff, 24.9.1961, Berol. Nos. 04959, 049427) Cherbourg (Leg. . . . 11.10.1836, ex herb. A. Le Jolis, No. 66

in K). Arromanche, Calvados (Leg. Lenormond, in K). Eastern end
of Calvados (Leg. Chauvin, Berol. No. 25059). Biarritz (Leg. L.
K. Rosenvinge, No. 655, 29.3.1885 ex herb. musei botanici Hau-
niensis, in K).
MALLORCA - Palma Nova (Leg. J. Gerloff, 15.2.1970, Berol. No.
25787). Paguera (Leg. J. Gerloff, 20.2.1970, Berol. No. 25803).
ENGLAND - Plymouth (Leg. J. Gatcombe, Berol. No. 10846 A. Leg.
I. Baswanar, Berol. No. 10846 B). Sidmouth and Torquay Bay
(Leg. . . . ex herb Wyatt, in K). HERBARIUM DE CANDOLLE GENEVE
(G), SWITZERLAND - There are three specimens in De Candolle's
herbarium (in G) labelled as *Ulva serrata* DC from (i) Coast of
Normandie, No. 358/6; Leg. Bouchet 1807. No. 358/1. (ii) Gulf
of Lyon, No. 358/3 B.

This species prefers calm conditions and has always been found
growing in very sheltered mid-littoral rocky pools along the
coast of Libya. Most probably this species grows throughout the
year.

Libyan specimens differ from those of the European coast of
Mediterranean Sea in possessing narrow fronds (1 cm broad)
whereas the latter are 4-6 cm broad and sometimes reach to
10 cm broad.

TAONIA ATOMARIA (Woodward) J. Ag. f. *CILIATA* (C. Agardh) comb.
Plate XI B, XII, XXXI, XXXII. Figs. 1, 24-28.

Basionym: *Zonaria atomaria* β *ciliata* C. Agardh 1820: 129; 1821: 129; 1823:
129; 1824: 265. Gray 1821: 341.

Synonyms: *Dictyota ciliata* Lamouroux 1809a: 41; 1809c: 331; 1813: 273. *Dictyota laciniata* Lamouroux 1809e: 41; 1809c: 331; 1812: 273. Duby 1830: 955.
Dictyota penicillata Lamouroux 1809a: 42; 1809: 331; 1813: 273. *Fucus pseu-
dociliatus* Lamouroux 1805: 41, t. 25, fig. 2. *Ulva serrata* De Candolle in
Lamarck et De Candolle 1805: 11. *Ulva fastigiata* Clemente 1807: 320. *Stypopodium atomaria* β *ciliatum* Kützing 1843: 341; 1849: 564; 1859: 24, t. 61,
fig. B. *Dictyota setosa* Duby 1830: 955.

Holdfast discoid to 1 cm diam. and felted, with a tereto-com-

37

pressed stipe to 5 mm long. Fronds tufted, erect to 22 cm high, cuneate below, flat above, di- or polychotomously branched into flat segments. Segments linear narrow, to 15 cm long, 5-10 mm broad, 150-240 μm thick, margin strongly dentate throughout as well as proliferated, composed of 6 or more layers of central cells covered on each side by a single layer of peripheral cells containing dense chromatophores. The number of layers of cells decreases towards the margins. The terminal parts or the young segments composed of three layers of cells, 40-70 μm thick. Peripheral cells radially elongated on either side of the surface. Fronds near the apices composed of 2 layers of cells. In surface view cells in longitudinal rows containing dense chromatophores. Apices of the segments splitting into further segments or lobes. Vegetative growth by means of marginal initials. Sori in zones or densely scattered. In old or mature condition zones not distinct but in young stage sori always in zones. Tetrasporangia ovoid, 80-120 x 40-70 μm, each producing 4 spores. Spores round, 80 x 80 μm across. Sporangial wall 2-layered, thick. Sex organs not observed.

Specimens examined: MOROCCO - Tanger (Leg. . . . 10.11.1879, as *Stypopodium ciliatum*, in K).

TUNISIA - Hammamat (Leg. J. Gerloff, 13.4.1968, Berol. No. 24519).

LIBYA - Tripoli, near Radio Station (Leg. M. Nizamuddin, 4.5. 1974). Tripoli, near Ender Hospital (Leg. M. Nizamuddin, 30.3. 1973, 15.4.1974, 22.4.1974). Cyrenaica (Leg. Hainmann, 1881).

CRETE - Stalis (Leg. J. Gerloff, 5.6.1974, Berol. No. 29369). Nordküste, near Candia (Leg. B. Watzl., 24.4.1914, Berol. No. 10853).

YUGOSLAVIA - Lissa Island, Dalmatia (Leg. V. Schiffner, 17.4. 1928, Berol. No. 25058).

SICILY - Palermo (Leg. L. Kny, 23.3.1870, Berol. No. 10845 A; 27.4.1870, Berol. No. 10852; 7.4.1871, Berol. No. 10856 B).

FRANCE - Roscoff (Leg. Runze, 18.9.1962, Berol. No. 17730). Arromanches, Calvados (Leg. Lenormand, in K).

HERBARIUM DE CANDOLLE IN GENEVE (G), SWITZERLAND - There are three specimens in De Candolle's herbarium (in G) labelled as *Ulva serrata* DC., (i) Gulf of Lyon, Nos. 358/3A; 358/4. (ii) Coast of Calvados (Leg. Lenormand, No. 358/5 as *Dictyota atomaria*

Taonia atomaria f. *ciliata* grows in mid-littoral region along the Libyan coast in association of *Dictyopteris tripolitana* and *Cystoseiras* in rough condition with strong wave actions.

T. atomaria f. *ciliata* resembles *T. atomaria* f. *atomaria* in habit but differs in possessing strong dentation, proliferations and in thick texture. Libyan specimens differ from those of the Roscoff (France) in having narrow segments. French specimens reach to 2.5 cm in breadth. Libyan specimens also differ structurally from that of Kützing (1859) in having more layers of central cells. It seems that Kützing had a young specimen at hand.

SPATOGLOSSUM Kützing 1843

Holdfast discoid and expanded giving rise to erect, foliaceous, flat fronds. Fronds, zonate, ecostate, irregularly sub-palmately to pinnately lobed. Vegetative growth by marginal initials. Thalli composed of several layers of cells in thickness with median layers of larger cells than the peripheral cells. Sori containing tetrasporangia scattered on both surfaces of the frond; oogonia dispersed over the surface; antheridia in small, scattered, slightly raised sori.

Type: *Spatoglossum solierii* (Chauvin) Kützing. (*Dictyota solierii* Chauvin).

Spatoglossum solierii is the only species so far reported from the Mediterranean coast.

SPATOGLOSSUM SOLIERII (Chauvin) Kützing
Plate XIV, XXXI. Fig. 3.

Basionym: *Dictyota solierii* Chauvin 1827: no. 2; 1842: 66. J. Agardh 1842: 37. Montagne 1836: 321.

Synonyms: *Taonia*(?) *atomaria* J. Agardh 1848: 103. Crouan et Crouan 1867: 169. *Laminaria padinipes* Bory in herb. nom. nud. Trevisan 1849: 449. *Dic-*

tyota latifolia Suhr 1839: 65 non J. Agardh 1894 non Kützing 1859. *Taonia solierii* Derbes 1856: 216.

J. Agardh 1894: 37. Ardissone 1883: 484. Ardrê 1969: 397. Bornet 1892: 229. Debray 1897: 48. Funk 1955: 51. Gayral 1958: 234. Giaccone 1969: 500. Hamel 1939: 335. Kützing 1843: 340; 1849: 560; 1849: 19, t. 46, fig. II. Rodrigues 1963: 56.

Holdfast discoid, large, expanded with numerous septate rhizoids. Stipe 1.5 cm long, cylindrico-terete, 2-3 mm broad. Fronds erect 10-40 cm high, flat to 2(-5) cm broad, surface smooth, upper parts deeply lobed alternately or irregularly or sub-dichotomously or palmately lobed or proliferated. Young fronds thin but the mature ones thick in texture, margin entire when young, lacerated or proliferated when old. Apices of lobes incised or fimbriated. Vegetative growth by means of marginal initials. Fronds faintly zonate in young stage only, composed of 5-7(-8) layers of flat cells. Peripheral cells contain dense chromatophores while others colourless. Tetrasporangia scattered and each produces 4 spores. Sexual organs not observed.

Specimens examined: MOROCCO - Tanger (Ex herb. Suhr, as *Dictyota latifolia* Suhr, in K selected as Lectotype).
LIBYA - Bengazi (Leg. M. Nizamuddin, 20.11.1977).
FRANCE - Marseille (Leg. L. Berner, 25.6.1929, Berol. No. 10834, on rocks in calm water in protected condition. Leg. Fr. Nees in Kützing 1843. Leg. Giraudy, 1894 as *Dictyota solierii* Chauv. in herb. identified by J. Agardh, in K).
BALEARES - (Leg. Biniscida, 29.5.1889 ex herb. J.J. Rodrigues, Algas de Menorca as *Spatoglossum solierii* Kg., in K).

Spatoglossum solierii is rather a rare species along the coast of the Mediterranean Sea. It is the first report from the Libyan coast. Specimens from the Gulf of Marseille are light yellow in colour, strongly lobed, surface smooth, texture thin, margins sparsely proliferated, basal part cylindrico-terete and adhering completely to paper. Libyan specimen, though a fragment, thick in texture, rough surface not adhering to paper. Suhr material in K to 4 cm broad, dark brown in colour, thick in texture and sporangia densely dispersed. Marseille specimens differ in texture from those of Libyan coast and Tanger but agree with that of *Spatoglossum areschougii* J. Ag.

DICTYOTA Lamouroux 1809

Fronds erect, brown or sometimes irridescent, attached by a small irregular fibrous holdfast; dichotomously branched into segments, each with a single apical cell; composed of a single layer of large central cells, regularly placed and covered on each side by a single layer of peripheral cells. Hairs in tufts on the surface of the frond tetrasporangia regularly scattered on the surface; antheridia and oogonia in sori.

Type: *Dictyota dichotoma* (Hudson) Lamouroux. (*Ulva dichotoma* Hudson).

Several species or varieties or forms belonging to the genus *Dictyota* were described or reported from the coast of the Mediterranean Sea but presently these are considered as vari-ables of one or the other. In the present text only three spe-cies, *D. dichotoma* (Hudson) Lamx., *D. linearis* (C. Ag.) Gre-ville and *D. pusilla* Lamx. are described.

KEY TO THE SPECIES

1. Segments to 5(-7) mm broad; sori scattered on the surface of the fronds
. . . . *D. dichotoma*
1. Segments to 2 mm broad, crypts or sori regularly arranged 2

2. Upper segments broad, flat, apices obtuse; sori transversely arranged
. . . . *D. linearis*
2. Upper segments thin filiform, cylindrical, apices narrowly acute; sori arranged medialy in a single row . . . . *D. pusilla*

DICTYOTA DICHOTOMA (Hudson) Lamouroux   Plate XV. Fig. A.

Basionym: *Ulva dichotoma* Hudson 1762: 476; 1778: 568; 1898: 568. Sowerby and Smith 1809: Pl. 774. De Candolle in Lamarck et De Candolle 1805: 11.

Synonyms: *Ulva punctata* Stackhouse 1897: 236. *Fucus zosteroides* Lamouroux 1805: 25, t. 22, fig. 3. *Zonaria dichotoma* C. Agardh 1817: XX; 1820: 133; 1821: 133; 1823: 133; 1824: 266. Gray 1821: 341. Greville 1824: 297. Suhr 1834: 724.

J. Agardh 1848: 92; 1882: 92; 1894: 67. Ardissone 1883: 478. Ardissone e Strafforello 1877: 163. Ardré 1969: 400. Batters 1902: 53. Børgesen 1926: 84; 1953: 15. Bornet 1892: 227. Crouan et Crouan 1867: 168. Debray 1893: 10; 1897: 47; 1899: 95. Decaisne 1841: 137. De Toni et Forti 1913: 14; 1914: 291. De Toni et Levi 1888: 478. Duby 1830: 954. Earle 1969: 157. Feldmann 1937: 317. Funk 1927: 362; 1955: 48. Gayral 1958: 218; 1966: 255. Gerloff und Geissler 1971: 750. Giaccone 1968: 223; 1969: 500. Gray 1867: 60. Greville 1830: 57, t. 10. Grunow in Heufler 1861: 425. Hamel 1939: 347. Harvey 1846-51: Pl. 103; 1849: 39; 1852: 109. Hauck 1885: 478. Hooker 1844: 284. Howe 1914: 71; 1920: 596. Kützing 1859: 5, t. 10, fig. I. Lamouroux 1809a: 42; 1809c: 331; 1813: 273. Langenbach 1873: 14. Le Jolis 1863: 97. Levring 1974: 31. Mazé et Schramm 1870-77: 118. Mazza 1903: 4; 1904: 121. Meneghini 1842: 224. Montagne 1840: 144; 1846: 30. Muschler 1910: 301. Nasr 1947: 79. Newton 1931: 212. Ollivier 1929: 116. Pampanini 1931: 35. Petersen 1918: 8. Piccone 1884: 26; 1892: 47. Reinbold 1898: 89. Reinke 1878: 3. Rodrigues 1963: 46. Schnetter 1976: 58. Taylor 1942: 58; 1960: 218; 1966: 355. Thuret 1878: 53. Zanardini 1847: 50. Zinova 1967: 140.

Fronds tufted, erect to 15 cm high, dichotomously branched into segments flat to 5(-7) mm broad, membranous, narrow near the terminal end, broadest near the dichotomy, angle of dichotomy broadly acute; margin entire; upper segments rarely spiral apices broadly acute to obtuse. Fronds composed of 3 layers of cells- a single layer of large central cells covered on each side by a single layer of peripheral cells containing dense chromatophores. Upper segments to 2 mm broad, 90-110 μm thick; cell walls thin (become wavy or dilated because of the fixatives). Lower segments to 5 mm broad, 120-150 μm thick but in this region cell walls thickened and lamellate to 20 μm thick. Sori irregularly scattered on both surfaces. Tetrasporangia ovoid, 140-170 x 80-140 μm; spores round to 100 μm across.

Specimens examined: MOROCCO - Tanger (Leg. Frl. von Weber. 188[

Berol. No. 106378. Leg. C. Karslädt, Berol. No. 10628).
TUNISIA - Gulf of Gabez (Leg. J. Gerloff, 9.8.1969, Berol. No.
25693).
LIBYA - Zanzoor (Leg. M. Nizamuddin, 10.7.1973, Leg. M. Gadeh,
10.7.1973). Andalus (Leg. M. Gadah, 10.7.1973). Tripoli, near
Marine Fishery Department (Leg. M. Nizamuddin, 30.4.1973).
Tripoli, near Ender Hospital (Leg. M. Gadeh, 30.3.1974). Tri-
poli (cf. Muschler, 1910). Leptis Magna (Leg. M. Nizamuddin,
15.6.1973. Leg. Bashir Fares, 28.5.1975). El-Khoms (Leg. M.
Nizamuddin, 23.5.1974). Tobruk (Leg. Vaccari, 1912). Bengazi
(Leg. Zanon, 1915).
GREECE - Syra Graecia Is. (Leg. Th. Liebetruth, 1870, Berol.
No. 10653, a fragment of *D. dichotoma* var. *minor* Kg.). Paleo-
kastritsa, Corfu (Leg. Kuhbier, May 1968, Berol. No. 24662, a
robust specimen from a depth of 12 m). Egina Is., near Agia
Marina (Leg. J. Gerloff, 16.4.1971, Berol. No. 26949). Cape
Sounion, west side (Leg. J. Gerloff, 22.5.1972, Berol. No.
28828, on rocks from 2-3 m depth).
YUGOSLAVIA - Guarnero, Loorana (June 1882, Berol. No. 10664).
Lovran, Istria (Leg. Zay, Flora Exs. Austro-Hungarica, Berol.
No. 10654B). Val di Bora, Rovinj (Leg. V. Schiffner, 2.8.1914,
Berol. No. 24703, from 2 m depth, as epiphyte on *Cystoseira
adriatica* Sauv. Leg. Vatova, 7.3.1932, Berol. No. 10655, from
2 m depth. Leg. J. Schiller, Feb. 1910, Berol. No. 10630, from
3 m depth). Rovinj, near the Institute (Leg. J. Gerloff, 7.6.
1961, Berol. No. 20036; 15.6.1961, Berol. No. 24664). Rovinj
(Leg. E. Lehmann, March 1903, Berol. No. 27948). San Giovanni,
Rovinj (Leg. V. Schiffner, Berol. No. 10660, from 4-5 m depth,
as epiphyte on *Cyst. adriatica* Sauv.). Pelagosa Is. (Leg. J.
Schiller, July 1912, Berol. No. 24706, from 6-25 m depth as
epiphyte on *Cystoseira* sp.; May 1912, Berol. No. 24692). Porto
Veglia, Veglia Is. (Leg. . . . 6.4.1915, Berol. No. 10663, as
f. *typica* and also f. *spiralis*. Leg. M. Lusina, 4.5.1915, 13.4.
1915, Berol. No. 10662). Pirano (Leg. P. Titius, Rabenhorst,
Algen Europas no. 1818, Berol. No. 10632). Ragusa, Lapad Peninsula (Leg. V. Schiffner, 4.4.1926, Berol. No. 10656, growing
in calm and protected habitat in ca. 1 m depth). Coast of Trappano, Sabbioncello Peninsula (Leg. F. Berger, end of September
1931, Berol. No. 24705, from 2 m depth).
ITALY - Porto Punta Franca, Trieste (Leg. J. Schiller, 1-3

43

April 1925, Berol. Nos. 25004, 25015). Sacchetta, Gulf of
Trieste (Leg. V. Schiffner, 12.8.1914, Berol. No. 10661B, from
0-3 m depth). Old Wavebreaker, Gulf of Trieste (Leg. V. Schiff-
ner, 12.8.1914, Berol. No. 10661C, from 1/2-1 m depth). On reef
before Miramar, Gulf of Trieste (Leg. V. Schiffner, 8.8.1914,
Berol. No. 10639). Trieste (Leg. Viv. mis. Cori, 20.11.1914,
Berol. No. 24702). Gulf of Naples (Leg. Masleno, Herb. Stock-
mayer ex herb. Grunow, Berol. No. 1064A). Port Area of Naples
(Leg. J. Schiller, 5-18 April 1925, Berol. Nos. 24700, 25008).
Porto Rende, Gulf of Naples (Leg. J. Brunnthaler, 7.4.1914,
Berol. No. 24704, as epiphyte on *Dictyopteris*. Leg. G. Funk,
10.3.1914, Berol. No. 24701, from 1 m depth. Lessina (. . . .
April 1844, as *Zonaria dichotoma*, Berol. No. 10640). Ischia,
near Naples (Leg. J. Kotiuzginin, 3.5.1870, Berol. No. 10641).
Genova (Leg. Martens, Berol. No. 10673).
SICILY - Arenella, Palermo (Leg. G. Langenbach, 27.5.1850,
Berol. No. 24693). Ficarazzi, Palermo (Leg. G. Langenbach, 14.
5.1870, Berol. No. 10643. Leg. L. Kny, 17.2.1870, Berol. No.
24696). Porto de Gran, Palermo (Leg. G. Langenbach, 24.1.1871,
Berol. No. 10642).
FRANCE - Cette (. . . ex herb. Bauschianum, Berol. No. 10631).
Roscoff, Channel de l'Ile Verte (Leg. Runze, 6.3.1962, Berol.
No. 17863).
HELOGLAND ISLAND - (Leg. J. Gerloff, Aug. 1938, Berol. No.
2389. Leg W. Panknin, 21.8.1935, Berol. No. 10638; 28.8.1936,
Berol. No. 07750; Aug. 1902, Berol. No. 27947. Leg. P. Kuckuck,
1.9.1905, Berol. No. 29865, 10649. Leg. J. Schiller, 15-30 Aug.
1924, Berol. No. 24691, from 1/2-3 m depth. Leg. Caspers, Sept.
1934, Berol. No. 24710. Leg. Pilger, Aug. 1911, Berol. No.
23288).
ENGLAND - Derby Port, Isle of Man (Leg. J. Gerloff, 27.8.1974,
Berol. No. 29545).
MALLORCA - Palma Nova (Leg. J. Gerloff, 25.2.1970, Berol. No.
25821). Magaluf (Leg. J. Gerloff, 17.2.1970, Berol. No. 25816;
19.2.1970, Berol. No. 25796). Paguera (Leg. J. Gerloff, 20.2.
1970, Berol. No. 25797, sublittoral).
EX MEDITERRANEAN COAST: Tanzania-Daressalam (Leg. Danke, Berol.
No. 10647). Red Sea (Leg. Von Frauenfeld, Herb. Stockmayer ex
herb. Grunow, Berol. No. 29760). Togo (Leg. Warucater, 1876,
Berol. No. 04380). Canary Islands - Puerto de la Cruz, Tene-

'iffe (Leg. J. Kohlmeyer, 10.12.1964, Berol. No. 27030). Punta
Hidalgo, Teneriffe (Leg. J. Gerloff, 21.10.1975, Berol. No.
29105, in calm places). St. Andres, Teneriffe (Leg. J. Gerloff,
.11.1973, Berol. No. 29120). North America - Pivers Island,
Beaufort, North Carolina (Leg. J. Kohlmeyer, 18.6.1962, Berol.
No. 23378). Town Marsh, Beaufort, North Carolina (Leg. J. Kohl-
meyer, 15.6.1963, Berol. No. 10634). Japan - Sunosaki (Leg.
Awa, April 1929, Berol. No. 10648). Vietnam - Nhatrang (Leg. E.
'. Dawson, 25.1.1953, Berol. No. 10634).

This species grows in the littoral regions in association of
aurencia spp., *Dictyopteris* sp. and *Enteromorpha* spp. along
the coast of Libya. The species prefers muddy rocky flat along
the Libyan coast. According to Petersen (1918) this species
was dredged from a depth of 75-90 m from Syrte Bay and 35 m
from Tripoli.

DICTYOTA DICHOTOMA (Hudson) Lamouroux var. *INTRICATA* (C. Agardh)
Greville   Plate XV B, XVI, XXXI, XXXII. Figs. 5, 36-41, 73-75.

Basionym: *Zonaria dichotoma* var. *intricata* C. Agardh 1820: 134; 1821: 134;
.823: 134; 1824: 266. Decaisne 1834: 10; 1841: 138. Suhr 1834: 724.

Synonyms: *Zonaria implexa* (Lamx.) C. Agardh 1817: XXI. *Fucus implexus* Desfontaines 1798: 423. *Dictyota implexa* (Desfont.) Lamouroux 1809a: 43; 1809c:
31; 1813: 273. J. Agardh 1842: 37; 1848: 92. Bory 1832: 321; 1838: 75.
Debray 1893: 10; 1897: 48; 1899: 95. Duby 1830: 955 (syn. p.p.). Decaisne
.841: 138. Giaccone 1968: 223; 1969: 500. Kützing 1859: 7, t. 14, fig. I.
Montagne 1840: 145; 1846: 30. Zinova 1967: 140. *Dictyota dichotoma* var. *implexa* (Desfont.) Gray 1821: 341. J. Agardh 1848: 92. Ardissone 1883: 478.
Bornet 1892: 227. Crouan et Crouan 1867: 168. Feldmann 1931: 216; 1937: 317.
Gerloff und Geissler 1971: 751. Hamel 1939: 349. Harvey-Gibson and Knight
.913: 306. Hauck 1885: 306. Le Jolis 1863: 98. Mazé et Schramm 1870-77: 118.
Newton 1931: 212.

Decaisne 1841: 138. Duby 1830: 954. Greville 1830: 58. Meneghini 1842: 227.
Papenfuss 1968: 33. Rodrigues 1963: 49. Zanardini 1847: 50.

Holdfast fibrous discoid, giving rise to many erect axes to 10

45

(-14) cm high, dichotomously branched into flat segments,
2-3 mm broad, thin. Fronds broadest (7 mm) near the dichotomy
of the basal parts. Upper parts dichotomously branched; segments linear, narrow, divaricate, twisted, entangled or inter-
wined, 450-520 μm (-1 mm) broad, to 50 μm thick. Terminal segments narrow, linear, to 3(-7) mm long, divaricate and apices
broadly acute. Angle of dichotomy acute in the lower parts.
Lower segments 820 μm broad and 80-130 μm thick. Fronds composed of 3 layers of cells. Tetrasporangia scattered on both
surfaces, 90-100 μm across.

Specimens examined: MOROCCO - Tanger (cf. Bornet, 1892).
LIBYA - Tripoli, near Marine Fishery Department (Leg. M. Niza-
muddin, 30.4.1974). Tripoli, near Ender Hospital (Leg. M.
Nizamuddin, 22.4.1974; 30.3.1974). Tripoli, along Sciare El-
Faateh (Leg. M. Nizamuddin, 15.6.1974). El-Khoms (Leg. M.
Nizamuddin, 12.4.1974).
GREECE - Attika, near Cape Sounion (Leg. J. Gerloff, 7.4.1971,
Berol. No. 27000). Killini, Peloponnisos (Leg. J. Gerloff, 15.
5.1975, Berol. 29717).
YUGOSLAVIA - Guarnero, Lovrano (June 1882, Berol. No. 10664A).
Coast of Trappano, Sabbioncello Peninsula (Leg. F. Berger, end
of September 1931, Berol. No. 24705 from 3 m depth). Rovinj,
near the port (Leg. J. Gerloff, 15.6.1961, Berol. Nos. 24665,
24666). Rovinj, near the Institute (Leg. J. Gerloff, 7.6.1961,
Berol. No. 20022, from 1.5 m depth). Val di Bora, Rovinj (Leg.
V. Schiffner, 2.8.1914, Berol. No. 25606, from 2 m depth; 17.7
1914, Berol. No. 10670, from 2 m depth). Rovinj (Berol. No.
10669). Orebic, Sabbioncello Peninsula (Leg. K.H. Rechinger,
April 1930, Berol. No. 25016). Valesca, near Veglia, Veglia Is
(Leg. M. Lusina, 29.4.1915, Berol. No. 25003). Lovran, Istria
(Leg. Zay, Flora Exs. Austro-Hungarica, Berol. No. 10654A).
ITALY - Frigs. (Leg. A. Grunow, Herb. Stockmayer, ex herb.
Grunow (Berol. No. 10646). Trieste, Port area (Leg. J. Schille:
15-17 June 1924, Berol. Nos. 25005, 25007, 25013). Punta Franc⊕
Gulf of Trieste (Leg. J. Schiller, 1-3 April 1925, Berol. No.
24667). Trieste (Leg. Viv. mis. Cori, 20.11.1914, Berol. No.
25012. Leg. J. Schiller, 20-25 Sept. 1924, Berol. No. 25017.
Leg. J. Schiller, 20-25 Sept. 1924, Berol. No. 25017. Leg.
Graeffe, February 1897, Berol. Nos. 10671; 25010). Tergestino

3ay, Adriatic Sea (Leg. F. Krasser, Krypt. Exs. no. 641, Berol. No. 10675). Port, Naples (Leg. Dufour, April 1959, Berol. No. 10674. Leg. J. Schiller, 5-18 April 1925, Berol. No. 24668. Leg. Seewald, Berol. No. 10665).

SICILY - Trapani (Leg. G. Langenbach, 5.9.1871, Berol. No. 10659A, 10672). Aspra, Palermo (Leg. G. Langenbach, 9.5.1870, 3erol. No. 24711). Palermo (Leg. E. Jauchen, 9.4.1913, Berol. No. 25611. Leg. L. Kny, 23.3.1870, Berol. No. 10657A).

FRANCE - Cherbourg (Leg. Le Jolis, September ? Berol. No. 25009). Hendaye (Leg. J. Kohlmeyer, 5.6.1962, Berol. No. 17519). SPAIN - Cadiz (Leg. Th. Liebetruth, Berol. No. 247000). ENGLAND - Treaeddur, Anglesey (Leg. J. Gerloff, 18.8.1974, 3erol. Nos. 29438, 19482).

HERBARIUM AGARDH, BOTANISKA MUSEET LUND (LD) - There are 5 specimens (48878-82) labelled as *Dictyota dichotoma* var. β *intricata* C. Ag. from different localities but none from Gadez, Atlantic coast of Morocco (the type locality). Mediterranean Sea (Leg. Lamouroux. LD. No. 48879 as *Fucus implexus* n.s.). Trieste (Leg. C. Agardh, LD. Nos. 48880; 48878 as *Zonaria dichotoma*). Madeira (Leg. Fr. Hall, LD. No. 48881 as *Ulva dichotoma* β *intricata* C. Ag.). Punta Delgada, Madeira (Leg. Fr. Hall, LD. No. 48882 as *Ulva dichotoma* β *intricata* C. Ag.). Specimen No. 48878 is selected as the lectotype in the absence of material from the type locality.

D. *dichotoma* var. *intricata* grows in the mid-littoral region in association of *Pterocladia capillacea* (Gmelin) Bornet et Thuret and *Enteromorpha linza* (L.) J. Ag. along the coast of Libya.

Var. *intricata* resembles var. *dichotoma* in habit and in general features but differs in possessing linear, slender, flexuous, twisted, entangled and divaricate parts.

DICTYOTA LINEARIS (C. Agardh) Greville
Plate XVII, XXXIII, XXXIV. Figs. 32-34, 35.

Basionym: *Zonaria linearis* C. Agardh 1817: XX; 1820: 134; 1821: 134; 1823: 134; 1824: 266 (excl. syn.).

Synonyms: *Dichophyllium lineare* (C. Ag.) Kützing 1843: 338. *D. angustissi-*
ma Kg. 1859: 10, t. 21, fig. II.

J. Agardh 1842: 37; 1848: 90; 1894: 77. Ardissone 1883: 481. Ardissone e
Strafforello 1877: 164. Børgesen 1914: 209; 1924: 22; 1926: 85. Bornet 1892
228. Debray 1897: 48. De Toni et Forti 1913: 14; 1914: 291. Earle 1969: 161
Feldmann 1931: 217; 1937: 317. Funk 1927: 362; 1955: 49. Gerloff und Geiss-
ler 1971: 751. Giaccone 1968: 223; 1969: 500. Greville 1830: Synopsis xliii
Hamel 1939: 349. Hauck 1885: 306. Kützing 1849: 556; 1859: 9, t. 21, fig.
II. Langenbach 1873: 14. Levring 1974: 32. Maze et Schramm 1870-77: 116.
Mazza 1903: 4; 1904: 123. Meneghini 1842: 221. Montagne 1838-1842: 65.
Muschler 1910: 301. Nizamuddin and Lehnberg 1970: 120. Pampanini 1931: 35.
Schnetter 1976: 63. Taylor 1942: 59; 1960: 219; 1966: 355. Zanardini 1847:
50. Zinova 1967: 140.

Fronds tufted, erect to 15 cm high, arising from a rhizoidal
discoid holdfast; dichotomously branched into thin, membranous
flat segments, 0.5-2 mm broad. Angle of dichotomy broad, obtuse. Upper segments linear, narrow but the terminal segments
almost broad to 300 μm thick; margin entire and apices obtuse.
Upper segments 350 μm broad, 100 μm thick having central cells
70 μm broad and 60 μm thick. Middle portions of the fronds flat
500-900 μm broad, 250 μm thick with central cells 60-70 μm
broad and to 170 μm thick. Lowermost portions of the fronds also flat, 1-2 mm broad, to 150 μm thick with central cells 70-
100 μm broad and to 100 μm thick. Peripheral cells 20-30 μm
thick, larger in the upper segments than those of the lower
ones. Cells of the basal parts as thick as broad, to 250 μm
with thickened walls. Sori, containing sporangia, transversely
arranged on the surface of the segments.

Specimens examined: MOROCCO - Bay of Tanger, Tanger (cf. Bornet
1892, July 1827).
TUNISIA - Between La Marsa and Sidi-Bou-Said (Leg. B. Schussnig
17.4.1924, Berol. No. 106682).
LIBYA - Tripoli (Leg. M. Gadeh, 15.4.1974. Leg. M. Nizamuddin,
22.4.1974. cf. A. Piccone, 1892, cf. Petersen, 26.7.1910, Sta.
No. 149). Gargaresh, Tripoli (Leg. M. Nizamuddin, 7.5.1973).
El-Khoms (Leg. M. Nizamuddin, 15.6.1973). Bengazi-Guiliana
(Leg. Zanon, 1915).

48

GREECE - Egina Is., near Agia Marina (Leg. J. Gerloff, 16.4.
1971, Berol. No. 27002). Cape Sounion (Leg. J. Gerloff, 22.5.
1972, Berol. No. 28827, on rocks, 2-3 m depth).
YUGOSLAVIA - Kostel Stari, near Trau (Leg. S. Varda, August
1925, Berol. No. 25023, 1 m depth as epiphyte on *Sargassum* sp.).
Vigna, Sabbioncella Peninsula (Leg. F. Berger, end of September
1931, Berol. No. 24699, from 1 m depth). Rovinj (Leg. J. Gerloff, 4.9.1965, Berol. No. 23704).
ITALY - Gulf of Trieste (Leg. J. Schiller, 20-25 September
1924, Berol. No. 24673). Riccione, near Rimini (Leg. Fr. Ebers-
bach, July 1924, Berol. No. 24675). Capri Island (Leg. E. Bauer,
August 1872, Berol. No. 10680). Come Sopra (Leg. Mazza, 1903).
SICILY - Mondello, Palermo (Leg. Mazza, 1904).

D. *linearis* as a mid- to sublittoral alga grows in the association of *Cystoseira compressa* (Esper) Gerloff and Nizam. and
Sargassum spp. On flat rocky platforms *D. linearis* grows in the
association of *Padina pavonia* (L.) Lamx. and *Enteromorpha* spp.
It was dredged from a depth of 5 m near Tripoli coast but it
has also been reported from a depth of 40 m from West Indies
(Børgesen, 1914).

D. *linearis* resembles *D. pusilla* Lamx. in habit but differs in
having broad terminal segments with obtuse apices and transversely arranged sori on the surface of the segments. In Rijks-
herbarium Leiden (L) No. 936.289-4 labelled *Dictyota angustissima* (Sonder Herb. ms.) Kg. from Corfu is a synonym of *D. linearis* but in the text type locality is Dalmatia (Leg. Sandri).
In the absence of any other material in Herb. Sonder, Melbourn.
No. 936, 289.4 is selected as the lectotype of *D. angustissima*
Kg.

C. Agardh (1817) accredited himself by publishing the manuscript name *Zonaria linearis* C. Ag. under the group "Lineis
fructum concentricis (Frons flabelliformis)" without any specific description. But later in 1820 and onwards he transferred
this species under another group, "Fronde dichotoma, segmentis
linearibus, fructibus sparsis" along with the specific description, "Fronde dichotoma integerrima, segmentis divaricatis linearibus rotundato-obtusis, zonis transversalibus". J. Agardh

(1848) elaborated the specific description in detail i.e.
"Fronde basi vix stuposa membranacea decomposito-dichotoma,
sinubus patentibus, segmentis angustissimis, margin integerrimis, ultimis sursum sublatioribus divaricatis, apicibus obtusis, aerolis rectangularibus longissimis, sori interruptis."
Further in his observations he explained the arrangement of
"sori i.e. sori per medium frondeum sparsi at (ob frondis elon-
gatione?) interrupti et discreti, quasi maculae tantum con-
spicui" but in 1894 he explained it differently i.e. "cellulas
fertiles supra medianum lineam segmentorum provenientes vidi".

The author has observed densely or sometimes sparsely
scattered sori on lower parts of the mature segments but in
the upper parts always transversely arranged (sometimes trans-
verse arrangement is wavy). Libyan specimens are in close con-
formity with the description given by C. Agardh. Type specimen
was not availabel for examination.

DICTYOTA PUSILLA Lamouroux   Plate XVIII A, XXXIII. Figs. 68-72.

Fronds tufted, flexuous, filiform, to 10 cm high, dichotomously or subdichotomously branched into filiform segments. Lower
parts of the frond proliferated, flat 850-1000 μm broad, 100-
250 μm thick. Upper segments cylindrico-terete, thin, 90-150 μm
thick, 200-350 μm broad. Terminal segments slender, filiform,
elongated, thin, 100-150 μm broad, 80-100 μm thick. Angle of
dichotomy acute, apices pointed and divaricate. Sori in a
single linear row along the central part of the segments.

Specimens examined: MOROCCO - Tanger (Leg. Dr. Th. Liebetruth,
1867. Berol. No. 10636 B, as *Dictyota implexa*).
LIBYA - Tripoli, near Radio Station (Leg. Dr. M. Nizamuddin,
7.5.1974). Tripoli, near Ender Hospital (Leg. M. Nizamuddin,
21.6.1971; 22.4.1974). Tripoli, Sciare El-Faateh (Leg. Dr. M.
Nizamuddin, 15.6.1974). Tajura (Leg. Dr. M. Nizamuddin, 10.3.
1974). Ghote El-Roman (Leg. Dr. M. Nizamuddin, 22.6.1974).
GREECE - Bay of Canone, Corfu (Leg. Dr. B. Watzl, 25.4.1914,
Berol. No. 24695, from 1 m depth, as *D. linearis* f. *divaricata*
Lamx.). Cape Sounion, west side (Leg. Dr. J. Gerloff, 20.5.

1972, Berol. No. 28831, as *D. linearis* f. *divaricata* Kg.).
Attika, near Cape Sounion (Leg. Dr. J. Gerloff, 7.4.1971,
Berol. No. 26951).

YUGOSLAVIA - North east of Bagnole Is., Rovinj (Leg. Dr. Krumbach, 20.6.1931, Berol. Nos. 10684, 24696, from 6 m depth, as
epiphyte on *Sargassum linifolium*, labelled as *D. linearis* f.
divaricata Lamx.). San Giovanni Island, Rovinj (Leg. Dr. V.
Schiffner, 28.7.1914, Berol. No. 24674, from 4-5 m depth, as
epiphyte on *Cystoseira adriatica*; Berol. No. 106834, as epiphyte on *Polysiphonia subulifera*). Zaate, east side (Leg. Dr.
A. Weiss, Herb. Stockmayer ex herb. Grunow, Berol. No. 24667,
as *D. divaricata* Lamx.; Berol. No. 27038, as epiphyte *Posidonia
oceanica* (L.) Delile.

ITALY - Lessina, Adriatic Sea (Leg. Col. W. Maccohio, 19.6.1845,
Berol. No. 24698). Livorno, Ligure Sea (Leg. Dr. F. Hauck, Herb.
Stockmayer ex herb. Grunow, Berol. No. 10681).

SICILIY - Palermo (Leg. Dr. L. Kny, 5.3.1870, Berol. No. 10657B,
as *D. dichotoma*).

FRANCE - Nice (Leg. Dr. Th. Liebetruth, 1864, Berol. No. 10636A,
as *D. intricata*). Cannes (Leg. Dr. A. Raphelis, August 1927,
Berol. No. 24697). Gulf of Marseille, Ille Ratoneau-Point
l'escourbidon (Leg. Dr. L. Berner, 11.7.1920, Berol. No. 24694,
as epiphyte on *Sargassum linifolium*, labelled as *D. linearis* f.
divaricata Kg. appears to be *D. pusilla* Kg. as there are marginal proliferations). Banyuls, Cape Béar (Leg. Runze, 4.9.
1962, Berol. No. 17915, dredged.

*Herbarium specimens in Museum Botanicum Universitatis Kielien-
is (K):* - Folder of *D. linearis* C. Ag.

Herbarium Sheet A - MOROCCO - Tanger (Leg. . . . 6.6.1879, as
Dictyota aequalis minor ex herb. Wollny). ADRIA - (Leg. . . .
Oct. 1878, as *Dictyota aequalis minor* , 28.9.1879, as *Dictyota
aequalis* ex herb. Wollny).

Herbarium Sheet B - YUGOSLAVIA - Syra (Leg. . . . ex herb. A.
le Jolis as *Dictyota linearis* Lamx.). Spalata (Leg. . . . 20.
.1878, as *Dictyota linearis*, ex herb. Wollny). ITALY - St.
Giuliano, Genova (Leg. A. Ardissone, as *Dictyota linearis*, as
Rabenhorst, Algen Europa's). FRANCE - Villefranche (Leg. J.
Agardh) 1841, as *Dictyota linearis*, ex herb. Suhr).

Herbarium Sheet C - YUGOSLAVIA - Zara (Leg. Sandri, as *Dictyota*

linearis Ag. No. 247). ITALY - Genova (Leg. Ardissone).
Herbarium Shett D - MOROCCO - Tanger (Leg. . . . 4.7.1879, as
Dictyota angustissima; 5.7.1879, as *Dictyota angustissima*; 20.3
1885, as *Dictyota angustissima*; 21.4.1879, as *Dictyota divaricata*; 4.7.1879, as *Dictyota fibrosa* ex herb. Wollny.
Specimens in Rijks-Herbarium Leiden (L): - No. 936,300 131 is
the lectotype of *Dictyota pusilla* Lamx. of Kg. where sori are
in linear arrangement but without tetrasporangia. Papillar
growth as shown in Tab. 25, fig. II, is nothing but the epiphytes. It is from Naples but the type is from *Catalania*. No.
936-289-3 is the type of *Dictyota fibrosa* Kg. from Naples is
slightly constricted and strongly divaricate terminal segment.
No. 936,289-100 labelled as *Dictyota divaricata* from Terraina
appears to be *Dictyota dichotoma* var. *intricata* (C. Ag.) Greville showing scattered sori. Okamura's (1913) *Dictyota linearis* appears to be *D. pusilla* Lamx. rather than *D. linearis*
(C. Ag.) Grev.

D. pusilla is a mid-littoral algae growing in rocky pools in
association of *Posidonia oceanica* (L.) Delile, *Dictyopteris* sp
and *Cystoseira discors* (L.) C. Ag. emend. Sauv. along the Libyan coast. *D. pusilla* also grows as epiphyte on other algae.
Along the European coast of the Mediterranean Sea this species
is a common epiphyte on *Sargassum linifolium* (Turn.) C. Ag. an
Cystoseira adriatica Sauv.

DILOPHUS J. Agardh 1882

Fronds tufted differentiated into stoloniferous and erect axes
Stolones terete, irregularly branched, attached by means of
rhizoids. Erect axes flat, flabellate or strap-shaped, ecostat
entire or denticulate, subdichotomously or alternately or irregularly branched into segments. Vegetative growth by means o
an apical cell. Fronds composed of 2 or more layers of large
central cells in thickness; in the lower parts or laterals or
proliferations, covered on each side by a single layer of peri
pheral cells but the upper parts or apical regions composed of
3 layers of cells as in *Dictyota*. Tetrasporangia single or mor
in sori.

Type: *Dilophus fasciola* (Roth) Howe (*Dictyota repens* J. Agardh: *Fucus fasciola* Roth).

Lamouroux (1809a) named *Dictyota fasciola* based on *Fucus fasciola* Roth and it is correctly written as *Dictyota fasciola* (Roth) Lamx. J. Agardh (1842) described a new taxa, *Dictyota repens* (from St. Hospice, Nice, French Mediterranean coast) which almost resembles in all respects to *D. fasciola* but he never considered *D. repens* and *D. fasciola* belonging to one and the same species. In 1882 he established a new genus *Dilophus* without assigning any holotype to the genus though he described six species of *Dilophus*. However, Howe (1914) was incorrect in saying that J. Agardh selected *D. repens* J. Ag. the type of his new genus *Dilophus*. In Herbarium Agardh, Botaniska Museet Lund, there are three specimens Nos. 49521-23 on a Herbarium sheet labelled as *Dictyota repens* J. Ag. None of these were selected as the type of the genus or the species. In the absence of any Holotype of the genus a Lectotype must be selected from the above mentioned specimens. No. 49523, d St. Hospice pr. Nizzam, Sept. 1841 legi (Herb. Agardh, und, Botaniska Museet Lund) is selected as Lectotype. Under the International Code of Botanical Nomenclature the specific epithet *repens* should be replaced by *fascicola*.

Hence the correct legal name of the type of *Dilophus* is *D. fasciola* (Roth) Howe and *Dictyota repens* J. Agardh should be conserved as the type of the genus *Dilophus*.

There is a great similarity between *Dictyota* Lamx. and *Dilophus* J. Ag. but the latter differs from the former in possessing stoloniferous axes, two or more layers of large central cells in the lower portions of the fronds, in texture and in attenuated base. The upper parts of the fronds of *Dilophus* possesses a single layer of large central cells similar to *Dictyota*.

From the Mediterranean Sea a number of species of *Dilophus* were described by previous authors but in fact only two species — *Dilophus fasciola* (Roth) Howe and *D. spiralis* (Montagne) Hamel are readily recognisable.

Terminal segments narrow, apices pointed . . . . *D. fasciola*
Terminal segments broad, spathulate, apices obtuse . . . . . . *D. spiralis*

DILOPHUS FASCIOLA (Roth) Howe
　　　　　　Plate XVIII B, XXXV, XXXVI. Figs. 52-62, 51.

Basionym: *Fucus fasciola* Roth 1797: 146, t. 7, fig. 1; 1800: 160.

Synonyms: *Fucus linearis* Forsskål 1775: 190, non *Fucus linearis* Hudson 1762
475. *Zonaria fasciola* (Roth) C. Agardh 1817: XXI; 1820: 136; 1821: 136;
1823: 136; 1824: 267. *Dictyota fasciola* (Roth) Lamouroux 1809a: 43; 1809c:
332; 1813: 273. J. Agardh 1842: 37; 1848: 89; 1882: 103; 1894: 78. Ardissone 1883: 480. Ardissone e Strafforello 1877: 164. Bornet 1892: 228. Bory
1832: 321; 1838: 75. Debray 1893: 10; 1897: 48. Duby 1830: 955. Funk 1927:
362. Greville 1830: Synop. 43. Grunow in Heufler 1861: 425. Harvey 1852:
108. Hauck 1885: 306. Kützing 1849: 555; 1859: 10, t. 22, fig. I. Maze et
Schramm 1870-77: 116. Mazza 1903: 4; 1904: 123. Meneghini 1842: 216. Muschler 1910: 301. Ollivier 1929: 117. Pampanini 1916: 290; 1931: 35. Piccone
1892: 48. Schiffner 1926: 306. Weber v. Bosse 1913: 186. Zanardini 1847:
50. *Dictyota linearis* Trevisan 1849: 452. ?*Dictyota abyssinica* Kützing 1859
9, t. 21, fig. III. *Dictyota acuminata* Kützing 1849: 555; 1859: 7, t. 15,
fig. III. *Dictyota acuta* Kützing 1845: 271; 1849: 555; 1859: 7, t. 13, fig.
I. *Dictyota aequalis* Kützing 1849: 555; 1859: 9, t. 21, fig. I. *Dictyota
aequalis* var. *minor* Kützing 1849: 555; 1859: 9, t. 20, fig. I. *Dictyota af-
finis* Kützing 1849: 554; 1859: 6, t. 12, fig. II. *Dictyota dichotoma* var.
acuta Chauvin in Duby 1830: 954. Crouan 1867: 168. *Dictyota ligulata* of
Vickers 1908: Pl. XIX. ?*Dictyota notarisii* Kützing 1859: 11, t. 25, fig.
III. *Dictyota striolata* Kützing 1847: 53; 1849: 554; 1859: 8, t. 17, fig.
II. Trevisan 1849: 453. *Dictyota verrucosa* Kützing 1859: 9, t. 19, fig. II.
Dilophus mediterraneus Schiffner 1931: 186. (incl. var. *crassus*) Gerloff
und Geissler 1971: 752.

Børgesen 1925: 82; 1938: 39. Feldmann 1931: 217; 1937: 309. Funk 1955: 50.
Gayral 1958: 224. Gerloff und Geissler 1971: 751. Giaccone 1968: 223; 1969
500. Hamel 1939: 351. Howe 1914: 72. Levring 1974: 31. Nasr 1947: 80.
Nizamuddin and Lehnberg 1970: 121. Papenfuss 1968: 33. Rodrigues 1963: 50.
Zinova 1967: 141.

Fronds differentiated into erect and stoloniferous axes attached to the substratum by means of rhizoids. Stoloniferous axes
terete to 500 μm broad and 250 μm thick. Erect axes tereto-flat,
3-4 mm broad, to 14 cm high, sub-dichotomously or alternately
or irregularly branched and strongly proliferated. Segment flat,
linear, narrow; margin entire; apices pointed, alternately or
irregularly proliferated. Proliferated laterals terete, flexuous, to 5 cm long, to 370 μm broad, to 250 μm thick. Basal part
of the erect axis terete to 400 μm broad, to 280 μm thick.
Young segments terete to 850 μm broad and 150 μm thick. Angle
of dichotomy acute and the first dichotomy generally at a distance of 3 cm. Stoloniferous axes, lower portions of the erect
axes and the proliferated laterals composed of many layers of
large central cells. Middle portion of the erect axes and the
flat segments comprises 4 layers of cells; 2 layers of large
central cells, 150-250 μm thick, and a single layer of peripheral cells, 30-50 μm thick on either side of the surface.
Apical parts of the erect axes and the young flat segments composed of 3 layers of cells — a single layer of large central
cells and covered on each side by a single layer of peripheral
cells. Proliferations develop from the surface of the erect
axes. These proliferations develop terete laterals with acute
apices possessing a single apical cell. These laterals gradually become compressed, 300-400 μm broad and 140-200 μm thick,
composed of many layers of large central cells but near the
terminal parts 2 layers of large central cells. Mature proliferated laterals 450-580 μm broad and 249-250 μm thick. Sori
numerous scattered on both surfaces of the segments. Tetrasporangia ovoid, 100-120 x 70-100 μm. Cryptostomata in a single
row on the proliferated laterals.

Specimens examined: ALGERIA - Cape Tizirine (Feldmann 1937.
Bornet (1892).

TUNISIA - Ain Ostar (Leg. J. Gerloff, 10.4.1968, Berol. No.
24522 as *Dilophus fasciola* var. *repens* (J. Ag.) Feldmann;
Berlol. No. 24520 as *Dictyota dichotoma* var. *implexa* (Desf.)
Gray). Carthage (Leg. B. Schussnig, 29.4.1924, Berol. No.
20754). Bay of Sfax (Leg. B. Schussnig, 24.4.1924, Berol. No.
25052).

LIBYA - Zanzoor (Leg. M. Gadeh, 10.7.1973). Gargaresh (Leg. M.

Nizamuddin, 24.5.1971). Tripoli, near Govt. Guest House (Leg. M. Nizamuddin, 12.5.1971, 15.6.1974). Tripoli, near Ender Hospital (Leg. M. Nizamuddin, 4.6.1974). Tripoli (cf. A. Piccone, 1892: 48). Ghote El-Roman (Leg. M. Nizamuddin, 25.5.1973). Bengazi-Sabri (Leg. Zanon, 1915).

ISRAEL - Mikhmoret (Leg. T. Rayss, 22.5.1961).

BLACK SEA - Novorassijks, Bay of Novorassijsk, Panticum Sea (Leg. Nina Vodjanitzviaja, 6.10.1925, Berol. No. 25048).

GREECE - Attika, near Cape Sounion (Leg. J. Gerloff, 7.4.1971, Berol. Nos. 26952, 26999, on rocky coast). Attika, near Vraona, rocky coast (Leg. J. Gerloff, 5.4.1971, Berol. No. 27004). Coast of Böotien, opposite Khalkis (Leg. K.H. Rechinger, 13.7. 1932, Berol. No. 25039, as *Dictyota fasciola* var. *divergens* Schiffner nom. nud.). Attika, near Vouliagmeni (Leg. K.H. Rechinger, 22.4.1927, Berol. No. 10756, as epiphyte on *Cystoseira* sp.). Killini, Peloponnisos (Leg. J. Gerloff, 15.5.1975, Berol. No. 29712, as *Dilophus spiralis* (Mont.) Hamel). Canone Beach, Corfu (Leg. B. Watzl, 25.4.1914, Berol. No. 25044, as epiphyte on *Cystoseira* sp. from 1 m depth Karystos, Euböa (Evia) (Leg. Fr. Werner, June 1937, Berol. No. 25053, as *Dilophus mediterraneus* Schiff. forma? as epiphyte on *Cystoseira* sp.). Cape Sounion, west side (Leg. J. Gerloff, 22.5.1972, Berol. No. 28811, as *Dilophus mediterraneus* Schiff.). Katápola Kykládes, Amorgós Island (Leg. K.H. Rechinger, June/July 1932, Berol. No. 10759, as *Dilophus mediterraneus* Schiff. var. *crassus* & *subteres* nom. nud.

CRETE - Tyvaki (Leg. K. Höfler, 10.4.1914, Berol. No. 25051, as epiphyte on *Cystoseira* sp.).

CYPRUS - (Leg. Unger, Herb. Agardh Lund, No. 49524, labelled as *Dilophus cypricus* Grunow ms. (= *Stypopodium flavum* Kg.?).

YUGOSLAVIA - Comisa, Lissa Island (Leg. V. Schiffner, 17.4. 1928, Berol. No. 10757, as *Dilophus fasciola* var. *angustissima* Kg., as epiphyte on *Cystoseira adriatica* Sauv. from 3 m depth) Spalato, Kasun Bay (Leg. V. Schiffner, 3.4.1928, Berol. No. 25042, as epiphyte on *Cystoseira barbata* (Gooden. et Woodw.) J. Ag. from 1 m depth). Lago Grande, Meleda Island (Leg. V. Schiffner, 2.4.1926, Berol. No. 25022, 25018 from 1 m depth; Berol. No. 25032, as *Dilophus mediterraneus* Schiff., as epiphyte on *Cystoseira pycnoclada* Schiff. ex Gerloff and Nizam. from 2-4 m depth). Between Makarska and Podgoro (Leg. F. Ber-

ger, June 1934, Berol. No. 25049 from 1 m depth). Ragusa, Lapad Peninsula (Leg. V. Schiffner, 4.4.1926, Berol. No. 25031).
Orebic, Sabbioncello Peninsula (Leg. F. Berger, 10.5.1930,
Berol. No. 10760, as epiphyte on *Cystoseira barbata* from 2-
5 m depth. This is the isotype of *Dilophus mediterraneus* Schiffner). Veglia City, Veglia Island (Leg. M. Lusina, 19.3.1915,
Berol. No. 10752, as *Dictyota fasciola* var. *acuta* Kg., as epiphyte on *Sargassum linifolium* (Turner) J. Ag.). San Sebastian,
Veglia Island (Leg. M. Lusina, 3.5.1915, Berol. No. 25043; 4.5.
1915, Berol. No. 25019; 16.4.1915, Berol. No. 25050, growing in
Enteromorpha zone). Scoglio Porere, Cape Promontore (Leg. Wett-
stein, February 1913, Berol. No. 10755, as epiphyte on *Cysto-
seira* sp.). Brioni, San Girolamo Is. (Leg. V. Schiffner, 30.7.
1914, Berol. No. 08611 from 10 m depth). Due Sorelle, Rovinj
(Leg. J. Gerloff, May 1970, Berol. No. 26363, 26364). Bagnole,
Rovinj (Leg. J. Gerloff, May 1970, Berol. No. 26926). Figarola,
Rovinj (Leg. J. Gerloff, May 1970, Berol. No. 26300). St. Ka-
tarina, Rovinj (Leg. J. Gerloff, 8.6.1961, Berol. No. 23717).
ITALY - Grignano, Trieste (Leg. G. Seefelder, 24.6.1913, Berol.
Nos. 10746, 25047 from 1 m depth). Portici, Gulf of Naples
(Leg. Mazza, 1903). San Nazzaro, Genova (Leg. A. Piccone, June
1860, Berol. No. 107 49).
SICILY - Arenella (Leg. L. Kny, 6.5.1870, Berol. Nos. 24698,
25020. Leg. G. Langenbach, 20.5.1870, Berol. No. 25021). Paler-
mo (Leg. Mazza, 1904).
FRANCE - Calanque de St. Estéve, Marseille (Leg. L. Berner, 21.
6.1930, Berol. No. 25002, as epiphyte on *Cystoseira* sp.) Sète
(Leg. . . . Berol. No. 1065, as *Zonaria dichotoma*).
HELGOLAND - (Leg. F. Bauer, 1849, Berol. No. 10747).
ENGLAND - Ramsgate (Leg. F. Bauer, 1863, Berol. No. 24707, as
Dictyota dichotoma ex herb. Soc. Bot. Lond.).
MALLORCA - Paguera (Leg. J. Gerloff, 20.2.1970, Berol. No.
25798).
U.S.A. - Key West, Florida (Leg. . . . ex herb. F.S. Collins.
Leg. C. Messina, Berol. Nos 10750, 10745).
MUSEUM BOTANICUM UNIVERSITATIS KIELIENSIS (K) - There is a spe-
cimen on Herharium Sheet No. A, Folder 2 (No. 16, labelled as
D. fasciola Roth) which certainly appears to belong to Suhr's
specimen on which the original description of *Dictyota cirrho-
sa* Suhr is based. Previously many authors considered this spe-

cies a synonym of *Dilophus fasciola* (Roth) Howe. Most probably
they considered *D. cirrhosa* Suhr a synonym of *Dil. fasciola* as
Suhr (1839) himself compared it with *Dictyota fasciola* (Roth)
Lamx. *Dictyota spiralis* Montagne (1846) was not known at the
time of his publication. Suhr's specimen in K externally appears to be similar to *Dilophus spiralis* (Mont.) Hamel but
anatomically differs from the latter as well as from *D. fasciola* in possessing 3 layers of cells, even from the basal part
of the frond. Hence *Dictyota cirrhosa* Suhr in K does not belong
to the genus *Dilophus* and is retained as *Dictyota cirrhosa* Suhr
and this specimen is selected as the lectotype of this species.
This specimen does not bear legit, dates and locality, but in
the text the locality is Tanger, Morocco.

Dilophus fasciola generally grows in mid- to sublittoral zones
in association of *Enteromorphos*, *Cladophoras* and *Cystoseiras*
along the Libyan coast. It was also dredged from 10-20 m depth.
C. Agardh (1820, 1821, 1823) reported the presence of a specimen of *Fucus spiralis* Forsskål (1775) a later homonym of *Fucus
linearis* Hudson (1762, 1778, 1798) = *Fucus ceranoides* var. *linearis* Batters in Herbarium Hornemann, Copenhagen and identified it as *Zonaria fasciola* (Roth) C. Ag. According to Børgeser
(1932) this specimen is lost. Therefore *Fucus fasciola* Roth is
the next available specific eipithet. So the correct name is
Dilophus fasciola (Roth) Howe.

There are variations within the species of *Dilophus fasciola*.
Some specimens are narrow, linear, thin; some are divergent
with acute apices; some posses acuminate apices; some possess
elongate, narrow terminal segments; some bear *Dictyota*-like
structure throughout except the base or the basal proliferated
laterals. These features overlap each other so such forms have
been merged into a single species — *Dilophus fasciola*. The
species which are based on the above characters are considered
as synonyms of *D. fasciola*. There is a variety — *D. fasciola*
var. *repens* which differs from *D. fasciola* var. *fasciola* in
possessing acute-obtuse apices and being smaller in size.

DILOPHUS FASCIOLA (Roth) Howe var. *REPENS* (J. Ag.) Feldmann
Plate XIX, XXXVIII. Figs. 67 A, B.

Basionym: *Dictyota repens* J. Agardh 1842: 38; 1848: 89. Bornet 1892: 228.
Debray 1897: 48. Kützing 1849: 554; 1859: 5, t. 9, fig. 1. Meneghini 1842:
219. Trevisan 1849: 453.

Synonyms: *Dictyota simplex* Kützing 1859: 5, t. 9, fig. II. *Dilophus repens*
J. Agardh 1882: 106; 1894: 86. Funk 1955: 50. Ollivier 1929: 117. Schiffner
1931: 188. *Dictyota fasciola* var. *repens* Ardissone 1883: 480. *Dictyota denti-*
culata Kützing 1849: 556; 1859: 12, t. 28, fig. I. Feldmann 1937: 312.
Hamel 1939: 351. Rodrigues 1963: 52. Zinova 1967: 143.

Fronds tufted, repent, forming mat-like habit, small to 5 cm
high to 1 mm broad, sub-dichotomously branched and entangled.
Lower parts proliferated. Segments cuneate below, flat above
with acute-obtuse apices. Upper segments, 215-610 µm broad, to
150 µm thick, composed of 3 layers of cells. Lower parts 500-
1500 µm broad, 200-280 µm thick, composed of 2 or more layers
of large central cells, covered on each side by a single layer
of peripheral cells. Terminal segments thick and angle of
dichotomy divaricate. Basal parts of the segments or the basal
proliferations terete or compressed, composed of more than 2
layers of large central cells. These proliferations generally
possess 5 or 6 large central cells but in some cases may be up
to 16 large cells but young proliferations flat above and composed of 3 layers of cells. Sori containing tetrasporangia
scattered on both surfaces of the fronds and each sporangia
producing 4 round spores.

Specimens examined: MOROCCO - Tanger (Leg. E. Bornet, 1892).
LIBYA - Tripoli, near Sciare El-Faateh (Leg. M. Nizamuddin, 15.
6.1974).
BLACK SEA - Akçakça, N. W. Anatolia (Leg. G. Wagenitz et H.J.
Beug, 23.9.1957, Berol. No. 27031).
FRANCE - Nice (Leg. J. Agardh, Sept. 1841, herb. Agardh Lund,
Botaniska Museet Lund, Nos. 49521-49523). J. Agardh (1842) did
not select any holotype for *Dictyota repens* (= *Dilophus repens*).
No. 49523 is selcted as the lectotype). Port Vendres, Dept.
Pyren. Orient (Leg. Frau E. Ronniger, 16.7.1927, Berol. No.
10753).

DILOPHUS SPIRALIS (Montagne) Hamel
                Plate XX, XXXVI, XXXVII. Figs. 42-50, 63-66.

Basionym: *Dictyota spiralis* Montagne 1846: 29; 1856: 397.

Synonyms: *Dictyota ligulata* Kützing 1847: 53; 1849: 554. 1859: 16, t. 18,
figs. a-d. J. Agardh 1848: 98; 1882: 95; 1894: 73. Børgesen 1926: 86. Bornet 1892: 228. Batters 1902: 53. Debray 1897: 48. Feldmann 1931: 216. Funk
1927: 363. Gerloff und Geissler 1971: 752. Giaccone 1968: 223; 1969: 500.
Grunow in Heufler 1861: 425. Mazza 1904: 122. Newton 1931: 213: Nizamuddin
and Lehnberg 1970: 121. Piccone 1901: 48, 54. Schiffner 1931: 186. Trevisan
1849: 453. *Dictyota dichotoma* var. *elongata* Crouan et Crouan 1867: 168.
Dictyota dichotoma var. *spiralis* Crouan et Crouan 1867: 168. Bornet 1892:
227. Debray 1893: 10. *Dictyota linearis* var. *spiralis* Ardissone 1883: 481.
Dictyota sibenicensis Zanardini in Kützing 1859: 5, t. 9, fig. IV. *Dilophus
ligulatus* (Kütz.) Feldmann 1937: 313. Funk 1955: 50. Gayral 220.

Ardrê 1969: 401. Gayral 1966: 257. Hamel 1939: 352. Zinova 1967: 144.

Fronds differentiated into stoloniferous and erect axes. Stoloniferous axes attached to the substratum by rhizoids and
axes giving rise to many erect, flat, subdichotomously branched
segments. Fronds to 12(-15) cm high, 2-4 mm broad, 100-150 µm
thick, subdichotomously or irregularly branched. Basal parts of
the erect axes attenuated functioning as stipe to 3 mm high,
compressed to 1 mm broad. Stoloniferous axes and laterals cy-
lindrico-compressed, to 500 µm broad and 140 µm thick. Terminal
segments flat, to 2 mm broad, spathulate, to 5 cm long, 140-
160 µm thick, composed of 3 layers of cells, margin entire;
apices broad or obtuse. Middle portion of the segments 150-
170 µm thick, composed of 4 layers of cells — 2 layers of
large central cells and a single layer of peripheral cells on
either side of the surface. Lower or basal portions of the
erect axes, composed of 4 layers of cells — 2 layers of large
central cells and a single layer of peripheral cells on either
side, but the margin with 3 layers of cells as in *Dictyota*.
Stoloniferous axes composed of many layers (more than 4) of
cells. Sori scattered or in patches along the median parts of
the segments. Tetrasporangia ovoid, 80-100 x 60-160 µm and
spores round, 100-120 µm across.

60

Specimens examined: MOROCCO - Tanger (cf. Schousboe in Bornet 1892: 228).

ALGERIA - La Calle (Leg. Durieu in Montagne 1846: 30). Cape Tizirine (cf. Feldmann, 1931: 216). Point-Pescade, near Alger (Leg. Bory in Montagne, 1846: 30, type locality).

TUNISIA - Raf Râf, rocky beach area (Leg. Mathieson and Menez, 9.9.1973).

LIBYA - Zanzoor (Leg. M. Gadeh, 10.7.1973). Andalus (Leg. M. Nizamuddin, 10.7.1973). Gargaresh (Leg. M. Nizamuddin, 24.5. 1971). Tripoli, near Govt. Guest House (Leg. M. Nizamuddin, 12. 5.1971). Ghote El-Roman (Leg. M. Nizamuddin, 22.6.1974, 26.4. 1975). El-Khoms (Leg. M. Nizamuddin, 23.5.1974). Leptis Magna (Leg. M. Nizamuddin, 22.5.1971).

GREECE - Attika, near Nea Makri (Leg. J. Gerloff, 11.4.1971, Berol. No. 27003).

CRETE - Stalis (Leg. J. Gerloff, 31.5.1974, Berol. No. 29344, on rock in about 1 m depth; 30.5.1974, Berol. No. 29392, on rocks above sublittoral; Berol. No. 20041, on rocks in 0.5 m depth; 11.6.1974, Berol. No. 29719, on rocks.

YUGOSLAVIA - Ragusa (Leg. V. Schiffner, 5.4.1926, Berol. No. 24672). Between Makarska and Podgoro (Leg. F. Berger, June 1934; Berol. No. 10666, as *Dictyota dichotoma* var. *attenuata* Kg. nom. nud.). Rovinj (Leg. Pittirsch, 28.9.1954, Berol. No. 10635, as *Dictyota dichotoma*).

ITALY - Cape di Posillipo, Gulf of Naples (Leg. G. Brunnthaler, 11.4.1914, Berol. No. 24671, as *Dictyota ligulata*).

SICILY - Cefalu (Leg. J. Kohlmeyer, 3.4.1958, Berol. No. 30795). Aspara, Palermo (Leg. G. Langenbach, 8.6.1870, Berol. No. 24691).

FRANCE - Banyuls sur Mer (Leg. J. Gerloff, 5.6.1962, Berol. No. 17558. Leg. J. Kohlmeyer, 24.5.1961, Berol. No. 23892, as *Dilophus fasciola*). Banyuls towards Cerbers (Leg. J. Kohlmeyer, 5. 6.1961, Berol. No. 17518, as *Dictyota dichotoma*). Cherbourg (Leg. L. Kny, 1867, Berol. No. 24699, as *Dictyota dichotoma*).

MALLORCA - Magaluf (Leg. J. Gerloff, 24.2.1970, Berol. No. 25794). Museum Botanicum Universitatis Kiliensis (K): - In Folder of *D. ligulata* there are specimens from Ex Herb. Holmes-(A). Lyme Regis (Leg. E.M. Holmes, Sept. 1889, as *Dictyota ligulata*. (B). Charmouth (Leg. E.M. Holmes, Sept. 1889 (C) England (Leg. E.M. Holmes as *Dictyota ligulata*).

D. spiralis is a mid-littoral alga and grows in association of *Hypneas* and *Cystoseiras* along the Libyan coast. It also grows in protected rocky pools. In some localities it grows in association of *Padinas*, *Ectocarpus* sp. and *Enteromorpha clathrata*. Along the Libyan coast this species grows all the year round. Antheridia and oogonia develop close to each other.

ABSTRACT

Along the coast of Libya seven genera and fourteen species belonging to the Dictyotales were reported. There are three species of *Dictyopteris* Lamx. - *D. membranacea* (Stackh.) Batters, *D. polypodioides* Lamx., *D. tripolitana* sp. nov.; three species of *Padina* Adanson - *P. pavonia* (L.) Lamx., *P. tenuis* Bory, *P. gymnospora* (Kg.) Vickers; three species of *Dictyota* Lamx. - *D. dichotoma* (Hudson) Lamx., *D. linearis* (C. Ag.) Grev., *D. pusilla* Lamx.; two species of *Dilophus* J. Ag. - *D. fasciola* Howe, *D. spiralis* (Mont.) Hamel; one species of *Taonia* J. Ag. - *T. atomaria* (Woodw.) J. Ag.; one species of *Spatoglossum* Kg. - *S. solierii* (Chauv.) Kg. and one species of *Zonaria* C. Ag. - *Z. flava* (Clem.) C. Ag. Distribution of these species has been described. Some nomenclatural and taxonomical problems related to these species were also discussed.

ACKNOWLEDGEMENT

The author wishes to thank Alexander von Humboldt-Stiftung, Bonn for financial assistance and to the Director, Botanischer Garten und Botanisches Museum, Berlin-Dahlem for providing facilities in completing this work. The author is also highly indebted to the Directors of the following institutions for the loan of the materials:
a) British Museum (NH) London (BM)
b) Universitetes Botaniska Museum, Lund (LD)
c) Conservatoire et Jardin Botaniques, Geneve (G)
d) Botanisches Institut, Universität Kiel (K)

e) Rijks Herbarium Leiden (L).

The author owes indebtedness to Mr. Muhammad Niaz for typing.

The author is also grateful to Prof. Dr. J. Gerloff for valu-
able advice and critically going through the manuscript.

REFERENCES

ADANSON, M. (1763) - Familles des Plantes. 2: 1 + (24) + [4] + 664. Paris.

AGARDH, C.A. (1817) - Synopsis algarum scandinaviae. xl + 135 pp. Lundae, Oct. 1817.

AGARDH, C.A. (1820) - Species algarum, rite cognitae cum synonymis, differentiis specificis et descriptionibus succinctis. 1(1): i-iv + 1-168, Lundae. Jan.-April 1820.

AGARDH, C.A. (1821) - Species algarum, rite cognitae cum synonymis, differentiis specificis et descriptionibus succinctis. 1(1): i-iv + 1-168, Gryphiswaldiae.

AGARDH, C.A. (1823) - Species algarum, rite cognitae cum synonymis, differentiis specificis et descriptionibus succinctis. 1(1): i-iv + 1-168, Gryphiswaldiae.

AGARDH, C.A. (1824) - Systema algarum. xxxviii + 312 pp. Lundae Oct. 1824.

AGARDH, J.G. (1841) - In Historiam Algarum Symbolae. Linnaea 15: 1-50, 443-457.

AGARDH, J.G. (1842) - Algae maris mediterranei et adriatici, observationes in diagnosin specierum et dispositionem generum. i-x + 164 pp. Parisiis. 9th April 1842.

AGARDH, J.G. (1848) - Species, Genera et Ordines Algarum-Fucoidearum. Vol. ., 363 pp. Lundae.

AGARDH, J.G. (1872) - Till Algernes Systematik. Lunds Univ. Årsskr. 9(8): .-71.

AGARDH, J.G. (1882) - Till Algernes Systematik. Lunds Univ. Årsskr. 1880-81, 7(4): 1-134, 3 Pls. Lund.

AGARDH, J.G. (1894) - Analecta Algologica . . . . Cont. I, Lunds Univer.
°
Arsskr. N.F. Afd. 29(9): 1-144, 2 tab. Lund.

ARDISSONE, F. (1883) - Phycologia Mediterranea - parte prima - Floridee.
i-x + 516 pp. Varese.

ARDISSONE, F. & J. STRAFFORELLO (1877) - Enumerazione delle Alghe di Liguria. 238 pp. Milano.

ARDRÉ, F. (1969-70) - Contribution a l'Étude des algues marines du Portugal
Portug. Acta Biol. Sér. B, 10(1-4): 137-555, 56 Pls.

ASKENASY, E. (1888) - Algen. In: Forschungsreise, S.M.S.M. 'Gazelle' in den
Jahren 1874 bis 1876 unter Kommando des Kapitän zur See Freiherrn von
Leinitz. Botanik, 4(2): 1-58, 12 taf. (Cover dates 1889 Berlin). Oct. 1888.

BATTERS, E.A.L. (1902) - A catalogue of the British marine algae. Jour. Bot
(Supplement) 40: 1-107. London.

BØRGESEN, F. (1914) - The marine algae of the Danish West Indies. 1(2): [4]
+ 159-226 + [2].Copenhagen.

BØRGESEN, F. (1924) - "Algae" In: Plants from Beata Island, St. Domingo
collected by C.H. OSTENFELD (Botanical results of the Dana-Expedition 1921-
22, No. 1). Dansk. Bot. Ark. 4(7): 14-36.

BØRGESEN, F. (1926) - Marine Algae from the Canary Islands especially from
Teneriffe and Gran Canaria II. Phaeophyceae. Det. Kgl. Dansk. Vidensk.
Selsk. Biol. Meddel. 6(2): 1-112.

BØRGESEN, F. (1930) - Some Indian green and brown algae especially from
the shores of the presidency of Bombay. Jour. Indian Bot. Soc. 9: 151-174.

BØRGESEN, F. (1932) - A revision of Forsskål's algae mentioned in Flora
aegyptiaco-arabica and found in his herbarium in the Botanical Museum of
the University of Copenhagen. Dansk. Bot. Ark. 8(2): 1-15, 1 Pl., 4 figs.

BØRGESEN, F. (1936) - Some marine algae from Ceylon. Ceylon Jour. Sci. Sect
A, Bot. 12(2): 57-96.

66

BØRGESEN, F. (1937a) - Contributions to a South Indian marine algal flora I. Jour. Ind. Bot. Soc. 16(1+2): 1-56.

BØRGESEN, F. (1937b) - Contributions to a South Indian marine algal flora II. Jour. Ind. Bot. Soc. 16(6): 311-357.

BØRGESEN, F. (1938a) - A list of marine algae from Bombay. Det. Kgl. Dansk. Vidensk. Selsk. Biol. Meddel. 12(2): 1-64, 10 Pls.

BØRGESEN, F. (1938b) - Sur une collection d'algues marines recueillies à une profondeur remarquable près des Iles Canaries. Rev. Algol. 11: 225-230.

BØRGESEN, F. (1939) - Marine algae from the Iranian Gulf especially from the innermost part near Bushire and the Island Kharg. Danish Sci. Invest. in Iran. 1: 1-141.

BØRGESEN, F. (1941) - Some marine algae from Mauritius II - Phaeophyceae. Det. Kgl. Dansk. Vidensk. Selsk. Biol. Meddel. 16(3): 1-81, 8 Pls.

BØRGESEN, F. (1948) - Some marine algae from Mauritius, additional lists to the Chlorophyceae and Phacophyceae. Det. Kgl. Dansk. Vidensk. Selsk. Biol. Meddel. 20(12): 1-55, 1 map, 2 Pls.

BØRGESEN, F. (1953) - Some marine algae from Mauritius - Additions to the plants previously published V. Det. Kgl. Dansk. Vidensk. Selsk. Biol. Meddel. 21(9): 1-62, 3 Pls.

BORNET, M.É. (1892) - Les Algues de P.K.A. SCHOUSBOE récoltées au Maroc & dans la Méditerranée de 1815 à 1829. Mém. Soc. Natn. Sci. Natur. et Math. 1er Cherbourg. 28: 165-376.

BORY, DE SAINT-VINCENT, M. (1827) - Padine - *Padina*. Bot. Crypt. (Hydro-phytes). Dict. Class. Hist. Nat., 12: 589-591. 18 Aug. 1827.

BORY, DE SAINT-VINCENT, M. (1828) - Cryptogâmie - [2] + 1-301. In: M.L.I. DUPERREY, Voyage Autore du Monde, exécuté par ordre du roi, sur la Covette de la majésté, LA COQUILLE, pendant les années 1822, 1823, 1824, et 1825, sous le ministre et conformément au instructions de S.E.M. le MARQUIS de CLERMONT-TONNERRE, ministre de la marines, et publié sous les auspices de on Excellence Mgr le Cte de Chabrol, ministere de la marine et des Colo-

nies. Botanique, par M.M. D'URVILLE, BORY De SAINT VINCENT et AD. BRONGIARI
Paris. [Imprimerie de Firmin Didot. Dec. 1828].

BORY, DE SAINT-VINCENT, M. (1832) - "Hydrophytes". In: Expédition Scientifique de Morée. Section des Sciences Physique-Botanique. 3(2): 316-334.
Paris and Strasbourg. Sept. 1832.

BORY, DE SAINT-VINCENT, M. (1838) - "Hydrophytes". In: Nouvelle Flore du
Péloponnèse et Cyclades par M. CHAUBARD. 74-78 pp. Paris and Strasbourg.

CANDOLLE, DE, A.P. (1818) - Regni vegetabilis systema naturae sive ordines,
genera et species, plantarum secundum methodi naturalis normas digistarum
et descriptarum. 1: 1-564, Paris.

CHAPMAN, V.J. (1963) - The marine algae of Jamaica. Bull. Inst. Jamaica
Sci. Ser. No. 12, 2: 1-201, Kingston.

CHAUVIN, F.J. (1827) - Exsiccatae. Algues de la Normandie. Fasc. 2, nos. 1-
100, Mém. Soc. Linn. Norm. Caen. (cf. Greville 1830: lxxv) Reliqiuae
Chauviniae was published and distributed around 1860 (cf. Flora, n.r. 18:
190-192: Bot. Zeit. 18: 109-111, 1860).

CHAUVIN, F.J. (1842) - Recherches sur l'Organisation, la fructification et
la classification de plusieurs genres d'algues, avec la description de que:
ques espèces in édites ou peu connues. Essai d'une repartition des polypieɩ
calciferes de Lamouroux dans la classe des Algues. 132 pp. Caen.

CLEMENTE, S. DE ROJAS (1804) - Essai sur les variétés de la Vigne, qui
végétent en Andalousie. Accedit appendix de Alguis Hispanicis. 1 Vol. in -
8. Paris (cf. De Candolle, 1818: 34).

CLEMENTE, S. DE ROJAS (1807) - Ensajo sobre las variedades de la vid comun
que vegetan en Andalucia, con un indice etimológico y tres listas de plantas en que se caracterizan varias especies nuevas. Lista tercera y ultima.
308-322 pp. con 2 laminas Madrid.

CROUAN, P.L. & H.M. CROUAN (1867) - Florule du Finistere - Plantes Cellulaires et Vasculaires. 262 pp., 31 pls. Brest.

DEBRAY, F. (1893) - Des algues marines et d'eau douce récoltées jusqu'a ce jour en Algérie. Bull. Sci. de la France et de la Belgique. 25: 1-19.

DEBRAY, F. (1897) - Catalogue des algues du Maroc, d'Algérie & de Tunisie. 78 pp. Alger.

DEBRAY, F. (1899) - Florule des algues marines du nord de la France. Bull. Sci. de la France et de la Belgique. 32: 1-193.

DECAISNE, M.J. (1834) - Énumération des plantes recueillies par BOVÉ dans les deux Arabies, La Palestine, la Syrée et l'Egypte. Ann. Sci. Nat. Bot. Sér. 2,2: 5-18.

DECAISNE, M.J. (1841) - Plantes de l'Arabie heureuse, recueillies par M.P. É. BOTTA. Archs. Mus. Hist. Nat. 2: 89-199, pls. 5-7, Paris.

DESFONTAINES, R.L. (1798) - "Fucus". In: Flora atlantica, sive historia plantarum quas in atlante, agro, Tunetano et algeriensi crescunt 2: 421-432, 16 pls. Parisiis. (Fucus - Part IV published Sept. 1798; Pls. 259, 260 published July 1799).

DE TONI, G.B. & A. FORTI (1913) - Contribution à la flore algologique de la Tripolitaine et de la Cyrénaique. Ann. Inst. Oceangr. 5(7): 1-56.

DE TONI, G.B. & A. FORTI (1914) - Plante Cellulari. "Algae". In: Plantae Tripolitanae (ed.) R. PAMPANINI. 289-305 pp., Firense.

DE TONI, G.B. & D. LEVI (1888) - Pugillo di alghe tripolitane. Acad. Lincei Roma Rendi. Sci. Fische, 4: 240-250.

DUBY, J.E. (1830) - "Algae". In: Botanicon Gallicum, sive synopsis plantarum in Flora Gallica descriptarum editio secund. Pars secunda, 935-994 pp., Paris May 1830.

EARLE, S.A. (1969) - Phaeophyta of the Eastern Gulf of Mexico. Phycologia, 7(2): 71-254.

ESPER, E.J.C. (1798) - Icones Fucorum, cum characteribus systematicis, Synonimis (sie) auctorum et descriptionibus novarum specierum 1(2): 55-126, 25-29a, 30-63. Nürnberg.

FELDMANN, J. (1931) - Contribution à la algologique marine de l'Algérie
les algues de Cherchell. Bull. de la Soc. Hist. Nat. de l'Afrique de Nord,
22: 179-254.

FELDMANN, J. (1937) - Algues marines de la Côte des Alberes-Phaeophycées.
Rev. Algol. 3: 243-335.

FORSSKÅL, P. (1775) - Flora Aegyptiaco-arabica, sive descriptiones plantarum, quas per Aegyptum inferiorem et Arabiam felicem, dextit, illustravit
edidit C. NIEBUHR. 32 + cxxvi + [2] 219 + [1] Hauniae.

FUNK, G. (1927) - Die Algenvegetation des Golfs von Neapel nach neueren
ökologischen Untersuchungen. Publ. d. Staz. Zool. di Napoli (Supplement).
7: 1-507, 20 Taf. Naples.

FUNK, G. (1955) - Beiträge zur Kenntnis der Meeresalgen von Neapel, zu-
gleich mikrophotographischer Atlas. Publs. d. Staz. Zool. di Napoli (Supple
ment). 25: 1-178 + i-x pp. 30 Taf. Naples.

GAILLON, B. (1828) - "Thalassiophytes". In: Dictionaire des Sciences Na-
turelles. 350-406 pp. Paris (Mai 1828).

GAYRAL, P. (1958) - Algues de la Côte Atlantique Marocaine. 523 pp. 152
Pls. Rabat.

GAYRAL, P. (1966) - Les Algues des Côtes Francaises (Manche et Atlantique).
632 pp. Paris.

GERLOFF, J. & U. GEISSLER (1971) - Eine revidierte Liste der Meeresalgen
Griechenlands. Nova Hedwigia, 22: (3+4): 721-793.

GIACCONE, G. (1968) - Raccolte di fitobenthos nel Mediterraneo Orientales.
Gior. Bot. Ital. 102(3): 217-228.

GIACCONE, G. (1969) - Raccolte di fitobenthos sulla Banchina Continentale
Italiana. Gior. Bot. Ital. 103: 485-514.

GMELIN, S.G. (1768) - Historia Fucorum [8] + 239 + 6 pp. 33 pls. Petropoli.

GRAY, S.F. (1821) - A natural arrangement of British plants, according to their relations to each other, as pointed out by JUSSIEU, DE CANDOLLE and C. BROWN including those cultivated for use; with an introduction to Botany, in which the terms newly introduced are explained; illustrated by figures. 1: i-xxvi + 1-824, pls. 1-21. Lonfon Oct.-Nov. 1821.

GRAY, S.O. (1867) - British Seaweeds - An introduction of the marine algae of the Great Britain, Iceland and the Channel Islands. i-xxiii + 312 pp., 16 pls. London.

GREVILLE, R.K. (1824) - Flora Edinensis or a description of plants growing near Edinburgh, arranged according to the Linnean System with a concise introduction to the Natural Orders of the Class Cryptogamia, and illustrated plates. i-lxxxi + 1-478 pp., 4 pls.

GREVILLE, R.K. (1830) - Algae britannicae or descriptions of the marine and other inarticulated plants of the British Islands belonging to the Order Algae; with plates illustrative of the genera. i-lxxxviii + 218 pp., 19 pls. Edinburgh Feb.-March 1830.

GRUNOW, A. (1861) - "Algas". In: Specimen Florae Cryptogamae septum insularum editum juxta Plantas Mazziarianas herbarii Heufleriani et speciatim quoad filices herbarii Tommasiniani. Verh. K. Zoo.-Bot. Gesellsch. 11: 416-430. Wien.

HAMEL, G. (1927) - Algues de France. Fasc. I, Nos. 1 à 50. Publié Rev. Algol. Paris.

HAMEL, G. (1931) - Algues de France. Fasc. III, Nos. 101 à 150. Publié Rev. Algol. Paris.

HAMEL, G. (1931-1939) - Phaeophycées de France. i-xlvi + 432 pp., 10 pls. Paris.

HARVEY-GIBSON, R.J. & M. KNIGHT (1913) - Reports on the marine biology of the Sudanese Red Sea. IX - Algae (Supplement). Jour. Linn. Soc. Bot. 41: 305-309. London.

HARVEY, W.H. (1846-1851) - Phycologia britannica. 1: i-lx pp., 186 pls. London.

HARVEY, W.H. (1849) - British Seaweeds - Manual of the British Marine Algae i-lii + 252 pp., 27 pls. London.

HARVEY, W.H. (1852) - Nereis boreali-americana. Part I. Melanospermae. 150 pp., 12 pls. Washington City.

HAUCK, F. (1885) - "Die Meeresalgen". In: Rabenhorst's Kryptogamen-Flora von Deutschland, Österreich und der Schweiz. 2: i-xxiii + 575 pp., 5 taf. Leipzig.

HAUCK, F. (1887) - Ueber einige von J.M. HILDEBRANDT im Rothen Meere und Indischen Ocean gesamelte Algen. Hedwigia, 26(2): 41-45.

HOOKER, W.J. (1844) - The English Flora of Sir JAMES EDWARD SMITH. Class XXIV. Cryptogamia. Vol. 1 (or Vol. 2 of Dr. HOOKER's the British Flora 1833 36). Part I comprising the Mosses, Hepaticae, Lichen, Characeae and algae. i-viii + 436 pp. London.

HOYT, W.D. (1921) - Marine algae of Beaufort, N.C. and adjacent regions. Bull. U.S. Bur. Fish. 1917-1918. 36: 367-556, 3 maps, pls. 84-100.

HOWE, M.A. (1914) - Marine algae of Peru. Mem. Torrey Bot. Club. 15: 1-185, 66 pls. 19 Sept. 1914.

HOWE, M.A. (1920) - "Algae". In: The Bahama Flora (ed.) N.L. BRITTON and C.F. MILLSPOUGH. 553-631 pp. New York.

HUDSON, W.G. (1762) - Flora anglica (edition 1, Jan.-June). [i] - viii, [1-8], 1-506, [1-22, ind.] pp., London.

HUDSON, W.G. (1778) - Flora anglica. (edition 2) vol. 1. [i-iii], 1-xxxviii [1, err], 1-334 pp. London. vol. 2. [i], 335-690 pp. London.

HUDSON, W.G. (1798) - Flora anglica (edition 3) [i-iii], i-iv, i-xxxii, 688 pp. London.

JOLY, A.B. (1965) - Flora marinha do littoral norte do estado de Sao Paulo e regoes Circunvizinhas. Boll. No. 294, Fas. Fil. Cienc. e letras da USP. Bot. 21: 1-393, 2 maps, 59 pls. Sao Paulo.

KÜTZING, F.T. (1843) - Phycologia generalis. i-xxxii + 458 pp., 79 tab. Leipzig. (14-16 Sept. 1843).

KÜTZING, F.T. (1845) - Phycologia germanica. 340 pp. Nordhausen. (14-16 Aug. 1845).

KÜTZING, F.T. (1847) - Diagnosen und Bemerkungen zu neuen oder kritischen Algen. Bot. Zeit. 5(3): 52-55. (22nd Jan. 1847).

KÜTZING, F.T. (23-24 July 1849) - Species algarum. i-vi + 922 pp. Lipsiae (exhibited in Oct. & Nov. 1849. In Flora No. 46 on 14.12.1849).

KÜTZING, F.T. (1859) - Tabulae phycologicae. 9: 1-100 tab. Lipsiae (1 Jan. - 9. March, 1859).

LAMARCK, J.B. DE & A.P. DE CANDOLLE (1805) - Florae francaise. Tome 2 (troisième edit.), i-xii + 600 pp. Paris.

LAMARCK, J.B. DE & A.P. DE CANDOLLE (1806) - Algae [4-13 pp.]. In: Synopsis Plantarum in Flora Gallica Descriptum xiv + 432 p. Parisiis.

LAMOUROUX, J.V.F. (1805) - Dissertations sur plusieurs espèces de *Fucus* peu connues ou nouvelles, avex leur description en latin et en francaise, Ier fascicule. A Agen de l'imprimerie de Raymond Noubel. An XIII, i-xxiv + 83 + 2 pp., 36 tab. Preface dated 1st May but announced as available in March 805.

LAMOUROUX, J.V.F. (1809a) - Exposition des Caracteres du genre *Dictyota* et tableau des espèces qu'il renferme. Jour. de Bot. 2: 38-44.

LAMOUROUX, J.V.F. (1809b) - Mémoire sur trois nouveaux genres de la famille des Algues marines. Jour. de Bot. 2: 129-135, 1 pl.

LAMOUROUX, J.V.F. (1809c) - Observations sur la physiologie des Algues marines, et description de cinq nouveaux genres de cette famille. Nouv. Bull. Sci. Soc. Philom. 1: 330-333, pl. 6, fig. 2. Paris. (Published in May, 809).

LAMOUROUX, J.V.F. (1813) - Essai sur les Genres de la famille des Thalassiophytes non articulées. Ann. Mus. Hist. Nat. 20: 267-293, 7-13 pls.

Paris. Présenté a l'Institut, dans la séance du 3 Fevrier, 1812.

LAMOUROUX, J.V.F. (1816) - Histoire des polypiers coralligênes flexibles vulgairement nommés zoophytes. i-lxxxiv + [1] + 1-559 + [1] pp., 19 pls. A Caen.

LANGENBACH, G. (1873) - Die Meeresalgen der Inseln Sizilien und Pantellaria i-viii + 1-24 pp. Berlin.

LANJOUW, J. (1952) - International Code of Botanical Nomenclature. 228 pp. Utrecht.

LE JOLIS, A. (1863) - Liste des Algues marines de Cherbourg. 168 pp. 6 pls. Paris.

LEVRING, T. (1974) - The marine algae of the Madeira. Boll. Mus. Munici. Funchal, 28(25): 5-111.

LINNAEUS, C. (1753) - Species plantarum 2: 561-1200. Holmiae.

LINNAEUS, C. (1759) - Systema naturae. Edition 10, 2: 1-2384. Holmiae.

LINNAEUS, C. (1763) - Species plantarum. Edition 2, 2: 785-1684. Holmiae.

LINNAEUS, C. (1764) - Species plantarum. Edition 3, 2: 785-1682, Holmiae.

LINNAEUS, C. (1767) - Systema naturae. Edition 12, 2: 1-736 + [8] pp. Holmiae.

LINNAEUS, C. (1770) - Systema naturae. Edition 13, 2: [2] + 1-736 pp. Vindobonae.

LINNAEUS, C. (1774) - Systema vegetabilium. Edition 13, 2: [8] + 844 pp. Gottingae et Gothae.

LINNAEUS, C. (1781) - Supplementum plantarum . . . . [16] + 467 pp. Holmiae.

LINNAEUS, C. (1792) - Systema naturae. Edition 13, 2(2): 885-1661. Holmiae.

MAZÉ, H. & A. SCHRAMM (1870-77) - Essai de classification des algues de la Guadeloupe. Facsimile edition. Ed. W. JUNK No. 8, (1905), Berlin NW. i-xix + 1-283 + [I-III] pp. 2e Edition. Basse-Terre (Guadeloupe).

MAZZA, A. (1903) - Flora marina del Golfo di Napoli. La Nuov. Notar. Ser. 4, 18: 1-17.

MAZZA, A. (1904) - Un manipolo di alghe marine della Sicilia. Part II. La Nuov. Notar, Ser. 15, 19: 115-149.

MENEGHINI, G. (1842) - Alghe italiane e dalmatiche. Fasc. III. fogli 11-16, 161-255 pp. Padova. Luglio 1842 (Reprint A. Asher & Co. N.V. 1970).

MISRA, J.N. (1966) - Phaeophyceae in India. I.C.A.R. Monographs on Alghae 10] + 203 pp. New Delhi.

MONTAGNE, J.F. CAM. (1836) - Notice sur les plantes Cryptogames récement découvertes en France, contenant aussi l'indication precise des localités de quelques aspèces les plus rares de la Flore française. Ann. Sci. Nat. : 321-339, avec une planche. Paris (December, 1836).

MONTAGNE, J.F. CAM. (1838-42) - Botanique-Plantes cellulaires. In: Histoire, Physique, Politique et Naturelle de l'Ile de Cuba par M. RAMON DE SAGRA. -x + 1-549, 15 taf. (Algae 1-104 pp.).

MONTAGNE, J.F. CAM. (1840) - Phytographia canariensis. Sectio ultima. Plantes cellulaires. In: Historia naturelle des Iles Canaries par PH. BARKER-WEBB et SABIN BETHELOT. 3(2): i-xv + 208, 9 pls. (I do not think that this work was published in 1840 as the Introduction dates 1841).

MONTAGNE, J.F. CAM. (1846) - "Phyceae". In: M.C. DURIEU DE MAISONNEURE, Exploration Scientifique de l'Algérie . . . Botanique-Flore d'Algérie . . . Cryptogamie. 197 pp., 16 pls. Paris.

MONTAGNE, J.F. CAM. (1856) - Sylloge generum specierumque Cryptogamarum. -xxiv + 498 pp. Paris.

RUSCHLER, R. (1910) - "Algae". In: E. DURAND et BARRAT, Flora Libycae Promus ou Catalogue Raisonné des Plantes de Tripolitaine. 293-313 pp. Genève.

NASR, A.H. (1847) - Synopsis of the marine algae of the Egyptian Red Sea coast. Fouad I Univ. Bull. Fac. Sci. 26: 1-156. Cairo.

NEWTON, L. (1931) - A handbook of the British Seaweeds. i-xiii + 478 pp. London.

NIZAMUDDIN, M. & W: LEHNBERG (1970) - Studies on the marine algae of Paros and Sikinos Islands, Greece. Bot. Marina, 13: 116-130.

NIZAMUDDIN, M. & S.M. SAIFULLAH (1967) - Studies on the marine algae of Karachi: *Dictyopteris* Lamouroux. Bot. Mar. 10(1/2): 167-179. (Mai).

NIZAMUDDIN, M., J.A. WEST & E.G. MENEZ (1979) - A List of Marine Algae from Libya. Bot. Mar. 22(7): 465-476.

OKAMURA, K. (1913) - Icones of Japanese algae. 3(2): 106-110 pls. Tokyo.

OLLIVIER, G. (1929) - Étude de la flore marine de la Côte d' Azur. Ann. Inst. Oceanogr. 7(3): 53-164. Paris.

OLTMANN, F. (1922) - Morphologie und Biologie der Algen. 2: 1-439. Jena.

PAMPANINI, R. (1931) - Prodromo della Flora Cirenaica. 40 pp.

PAPENFUSS, G.F. (1968) - A history, catalogue and bibliography of Red Sea benthic algae. Isr. Jour. Bot. 17(1-2): 1-118, 1 map.

PETERSEN, H.E. (1918) - Report on the Danish Oceanographical Expedition 1908-1910 to the Mediterranean and adjacent seas. Biol. K_3, 2(5): 1-20.

PICCONE, A. (1884) - Crociera del 'Corsaro' alle isole Madera e Canarie del Capitano Enrice d'Albertis. 60 pp., 1 pl. Genova.

PICCONE, A. (1892) - Alghe della Cirenaica. Ann. Roy. Ist. Bot. 5(2): 45-52. Roma.

PICCONE, A. (1901) - Noterelle ficologiche XI-XV. La Nuov. Notar., 12: 45-58.

REINBOLD, T. (1898) - Meeresalgen von der Insel Rhodos. Hedwigia, 37: 88-9

REINKE, J. (1878) - Entwicklungsgeschichtliche Untersuchungen über die Dictyotaceen des Golfs von Neapel. Nov. Act. Leop.-Carol. Acad. 40(1): 1-48, 7 tab.

ROBINSON, W. (1932) - Observations on the development of *Taonia atomaria* Ag. Ann. Bot. 46: 113-120. Pl. VI.

RODRIGUES, J.E. DE MESQUITA (1963) - Contibucãe para o Conhecimento das Phaeophyceae da Costa Portuguesa. Mem. da Soc. Bot. Inst. Bot. da Univ. de Coimbra. 16: 1-124, 18 pls. Coimbra.

ROTH, A.W. (1797) - Catalecta botanica, quibus plantae novae et minus cognitae describuntur atque illustrantur. Fasc. I, 146-387 pp. 29 tab. Lipsiae.

ROTH, A.W. (1800) - Catalecta botanica, quibus plantae novae et minus cognitae describuntur atque illustrantur. Fasc. II, 158-250 pp. Lipsiae.

SCHIFFNER, V. (1926) - Beiträge zur Kenntnis der Meeresalgen. Hedwigia, 66: 293-320.

SCHIFFNER, V. (1931) - Neue und bemerkenswerte Meeresalgen. Hedwigia, 71: 139-205.

SCHNETTER, R. (1976) - Marine Algen der karibischen Küsten von Kolumbien. I - Phaeophyceae. Bibl. Phycol. 24: 1-125. (J. Cramer).

SEGAWA, S. (1960) - Coloured illustrations of the sea weeds of Japan. I-XVIII + 1-175 pp. Osaka, Japan.

SETCHELL, W.A. (1926) - Tahitian algae, collected by W.A. SETCHELL, C.B. SETCHELL and H.E. PARKS. Univ. Calif. Publs. Bot. 12(5): 61-142. Pls. 7-22.

SILVA, P.C. (1952) - A review of nomenclatural conservation in the algae from the point of view of the type method. Univ. Calif. Publs. Bot. 25(4): 241-324.

SIMONS, R.H. (1964) - Notes on the species of *Zonaria* in South Africa. Bothalia 8(2): 195-220.

SRINIVASAN, K.S. (1969) - Phycologia indica (Icones of Indian marine algae)

1: i-xix + 1-52 pp., 51 pls. Bot. Survey of India. Calcutta.

SRINIVASAN, K.S. (1973) - Phycologia Indica (Icones of Indian marine algae) 2: i-xvii + 1-60 pp., 54 pls. Bot. Survey of Calcutta.

STACKHOUSE, J. (1795) - Neries britannica; containing all the species of Fuci, natives of the British coasts: with a description in English and Latin, and plates coloured from nature. Fasc. 1, 1: 1-30. Bath.

STACKHOUSE, J. (1797) - Description of *Ulva punctata*. Trans. Linn. Soc. 3: 236-237 (Read on May 5, 1795).

SOWERBY, J. & J.E. SMITH (1795-1812) - Icones Thalassiophycorum. In: Anglica botanica. Tome 1, 114 pp., 114 pls.

STAFLEU, F.A. (1972) - International Code of Botantical Nomenclature. 426 pp. Utrecht.

STAFLEU, F.A. (1978) - International Code of Botanical Nomenclature. i-xiv + 457 pp. Utrecht.

STAFLEU, F.A. & R.S. COWAN (1976) - Taxonomic Literature. Vol. 1: A-G. 1136 pp. 2nd edition. Utrecht.

SUHR, J.N. von (1834) - Uebersicht der Algen, welche von HRN. ECKLON an der südafrikanischen Küste gefunden worden sind. Flora, 17(2): 721-735. Regensburg.

SUHR, J.N. von (1839) - Beiträge zur Algenkunde. Flora, 22(1): 65-75, 1 Tab. Regensburg.

TAYLOR, W. R. (1942) - Caribbean marine algae of the Allan Hancock Expeditions. Univ. South Calif. Publs. Allan Hancock Atlantic Expedition Report No. 2, 193 pp.

TAYLOR, W.R. (1960) - Marine algae of the eastern tropical and subtropical coasts of the Americas. Univ. Mich. Stud. Sci. Ser. 21 (edition 1), lx + [1]-870 pp. Ann Arbor.

TAYLOR, W.R. (1966) - Records of Asian and Western Pacific Marine Algae, particularly algae from Indonesia and the Philippines. Pacific Science. 20(3): 342-359.

THIVY, E. (1959) - On the morphology of the gametophytic generation of *Padina gymnospora* (Kuetz.) Vickers. Jour. Mar. Biol. Assoc. India. 1(1): 69-76.

THURET, M.G. (1878) - Études Phycologiques-Analyses d'algues marines. 101 pp., 51 pls. (Réimp. 1966). Paris.

TREVISAN, V.B.A. Comte de (1849) - Dictyoteis Adumbratio. Linnaea, 22(4): 421-464. August 1849.

VICKERS, A. (1905) - Liste des Algues marines de la Barbade. Ann. Sci. Nat. Bot. Ser. 9, 1: 45-66.

VICKERS, A. (1908) - Phycologia barbadensis - Iconographie des algues marines récoltée a l'île Barbade (Antilles): Text by M.H. SHAW. Partie II - Phaeophyceae, 33-44 pp., 34 pls. Paris.

WEBER, F. & D.M.H. MOHR (1805) - Einige Worte über unsere bisherigen, hauptsächlich karpologischen Zergliederungen von kryptogamischen Seegewächsen. Beitr. zur Naturk. 1: 204-329. Kiel.

WEBER VAN BOSSE, A. (1913) - Liste des algues du Siboga. I - Myxophyceae, Chlorophyceae, Phaeophyceae, avec la concours de M.TH. REINBOLD. Monographie, 59a: 1-186, 5 pls. Leiden.

WEBER VAN BOSSE, A. (1928) - Liste des algues du Siboga. Siboga-Expeditie. 59d, 4(3): 393-533. Pls. 11-16. Leiden.

WILDEMAN, É. de (1910) - Actes du IIIme Congrès International de Botanique. Vol. 1. Comptes-rendus séances, excursions etc. 385 pp. Bruxelles.

WOODWARD, TH.J. (1797) - Observations upon the generic character of *Ulva*, with descriptions of some new species. Trans. Linn. Soc. 3: 46-58. London.

ZANARDINI, G. (1847) - "Prospetto Alghe marine". In: Prospetto della Veneta. 39-53 pp. Venezia.

ZINOVA, A.D. (1967) - Opredeli sel yelenyen buryeh i krasnyeh vodoraslej Juznyeh morey SSSR. 398 pp. Moscow. (Diagnoses of green, brown and red algae of the Southern Seas of USSR).

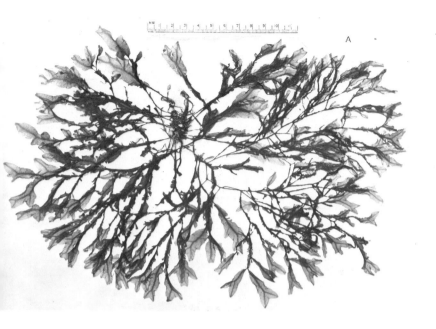

Plate I. *Dictyopteris membranacea* (Stackhouse) Batters: (A) Habit of the
frond (Berol. No. 10707). (B) Scattered sori on either side of the midrib.

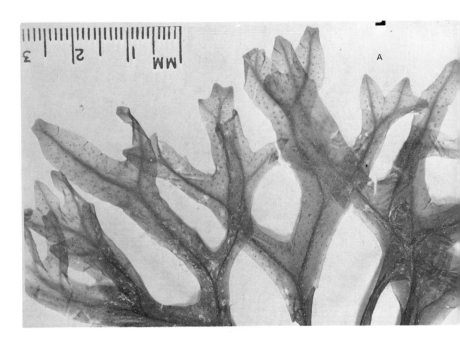

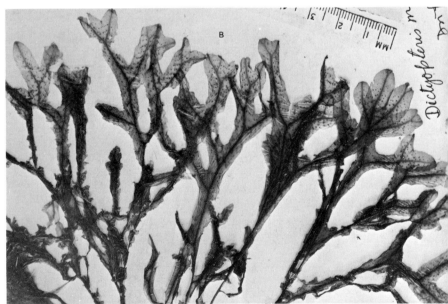

Plate II. *Dictyopteris membranacea:* (A) Arrangement of sori on either side of the midrib. *Dictyopteris tripolitana* sp. nov. (B) Transversally-obliquely arranged sori on either side of the midrib (No. 0512).

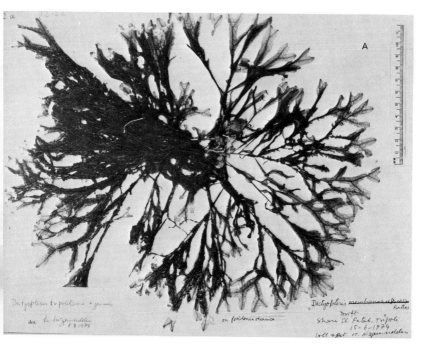

Plate III. *Dictyopteris tripolitana:* (A) Habit of the frond. *Dictyopteris polypodioides* Lamouroux: (B) Habit of the frond (No. 1245).

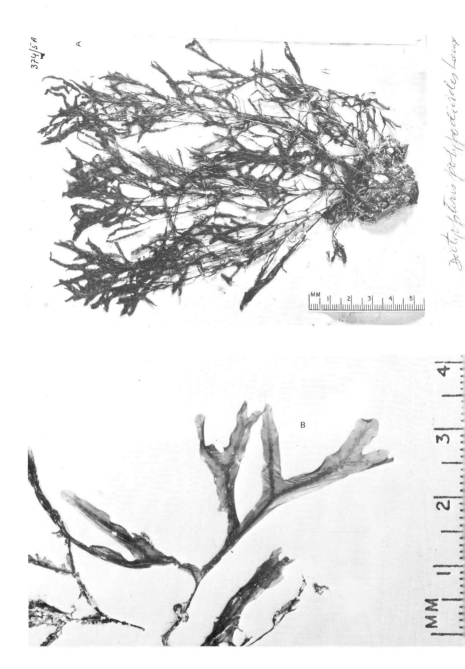

Plate IV. *Dictyopteris polypodioides* (= *Ulva polypodioides* DC): (A) Habit of the frond (No. 3745A in G). *Dictyopteris polypodioides:* (B) Arrangement of sori in patches along the midrib.

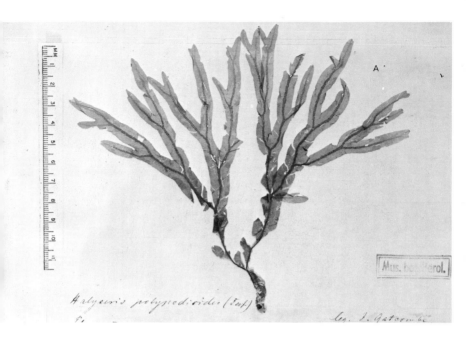

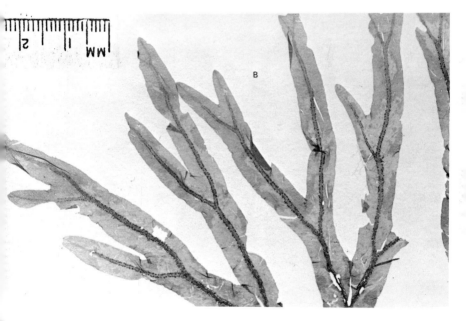

Plate V. *Dictyopteris polypodioides:* (A) Habit of the frond (Plymouth Leg.
C. Gatcombe, Berol. No. 10731). (B) Upper part of the frond showing arrangement of sori along the midrib.

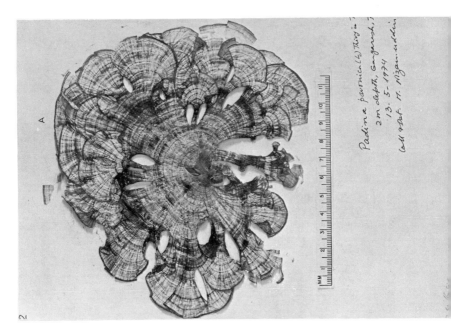

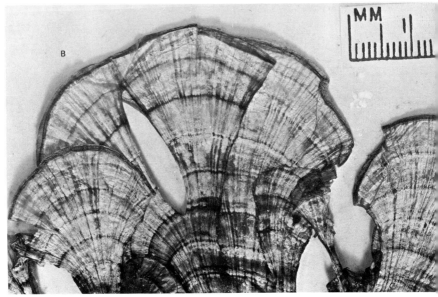

Plate VI. *Padina pavonia* (L.) Lamx.: (A) Habit of the frond. (B) Upper sur
face showing arrangement of sori.

86

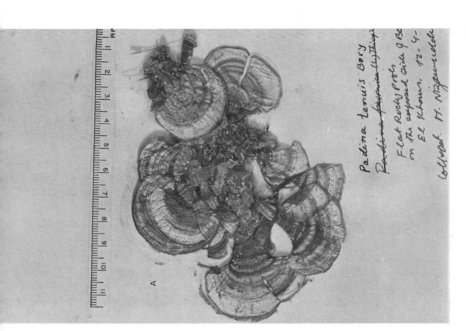

Plate VII. *Padina tenuis* Bory: (A) Habit of the frond. (B) Upper surface
showing arrangement of sori.

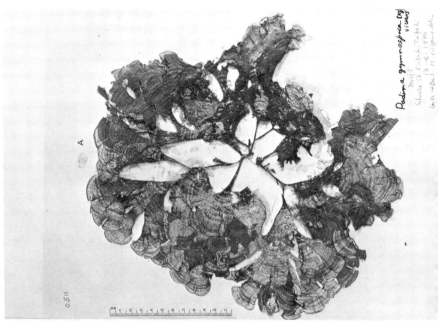

Plate VIII. *Padina gymnospora* (Kg.) Vickers: (A) Habit of the frond. (B)
Upper surface showing arrangement of sori.

88

Plate IX. *Zonaria flava* (Clemente) C. Agardh: (A) Habit of the frond (Benghasi, Drift No. 0544). (B) Habit of the frond (Gulf of Naples, No. 10868B).

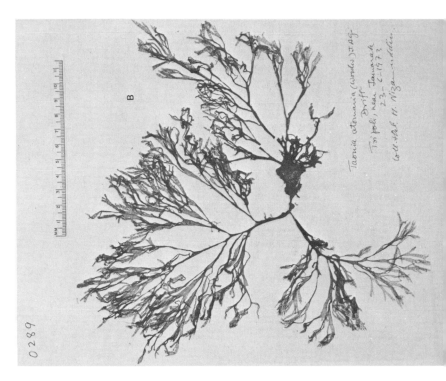

Plate X. *Zonaria flava:* (A) Habit of the frond (Teneriffe, Berol. No. 29141). *Taonia atomaria* (Woodward) J. Agardh: (B) Habit of the frond.

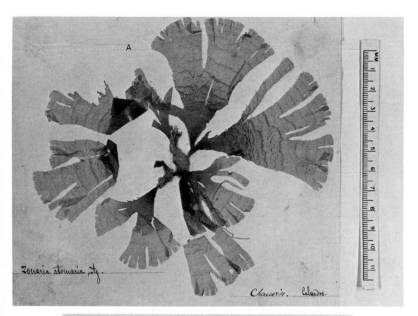

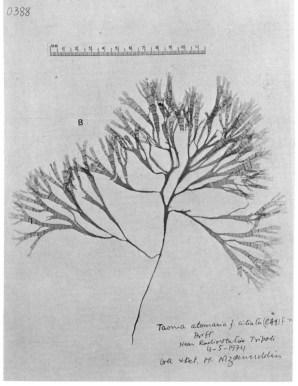

Plate XI. *Taonia atomaria* f. *atomaria:* (A) Habit of the frond (Calvados, Berol. No. 25059). *Taonia atomaria* f. *ciliata:* (B) Habit of the frond (Tripoli).

6 cm

le centionnie

Dictyota — Atomaria — Grev.

Arromanches (Calva

Plate XII. *Taonia atomaria* f. *ciliata:* (A) Habit of the frond (Arromanche, Calvados in K).

Herb. G. THURET.

Taonia atomaria. J. Ag.

S.ᵗ Vaast! 2 Octobre 1853.

Donné par Mᵣ **Ed. Bornet.**

late XIII. *Taonia atomaria* f. *atomaria*: (A) Habit of the frond (St. Vaast
n K).

93

Schirner, Algae marinae № 689.

Spatoglossum Solierii (Chauv.) Ktz.
planta rara;
Mittelmeer: Golfe de Marseille;
Chateau d'If. Felsen in ruhigen Wasser
an schattigen Stelle,
25 6.1920 leg. I. Huber u. Schirner

A

Spatoglossum solierii (Chauv) Kg.
Diff
Bengasi. 20-11-1977

B

Plate XIV. *Spatoglossum solierii* (Chauvin) Kützing: (A) Habit of the frond
(Gulf of Marseille. Berol. No. 10834). (B) A fragment from Bengasi, Libya

94

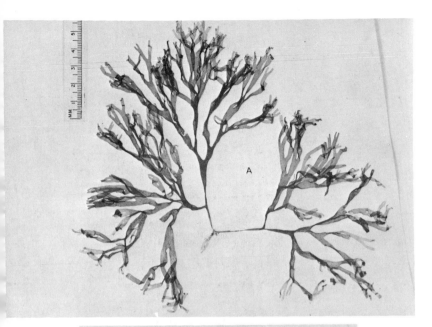

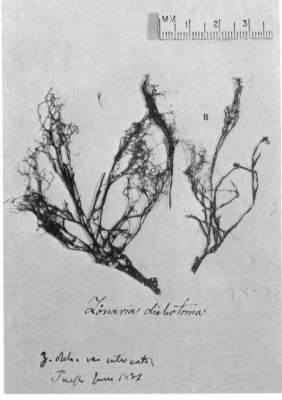

Zonaria dichotoma

Z. dich. var. intricata
Juel June 1831

Plate XV. *Dictyota dichotoma* (Hudson) Lamouroux: (A) Habitat of the frond
(Tripoli). *Dictyota dichotoma* var. *intricata* (C. Agardh) Greville: (B)
Madeira, Lectotype No. 48878 in L). Habit.

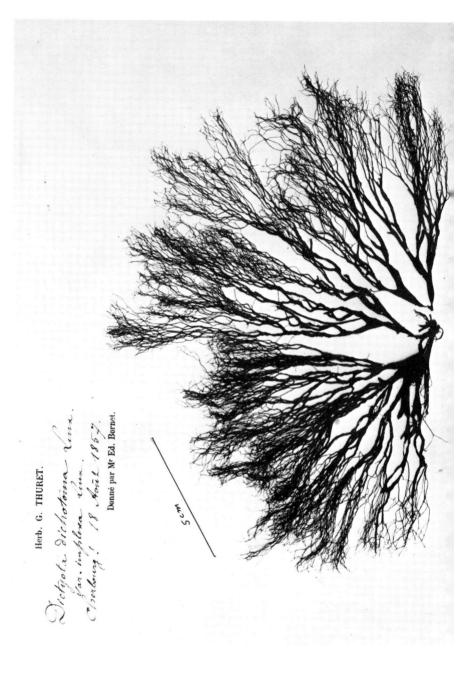

5 cm

Plate XVI. *Dictyota dichotoma* var. *intricata*: (A) Habit of the frond.

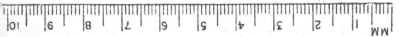

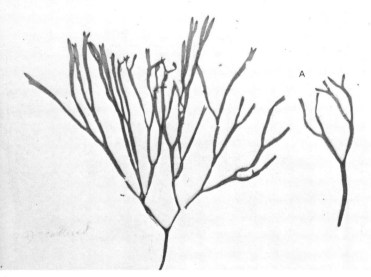

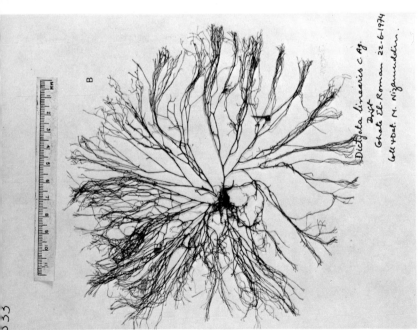

Plate XVII. *Dictyota linearis* (C. Agardh) Greville: (A) Habit of the frond
(Adriatic Sea). *Dictyota linearis* (C. Agardh) Greville: (B) Habit of the
frond (Ghote Roman, Tripoli).

97

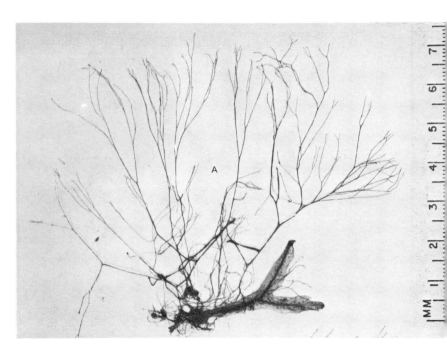

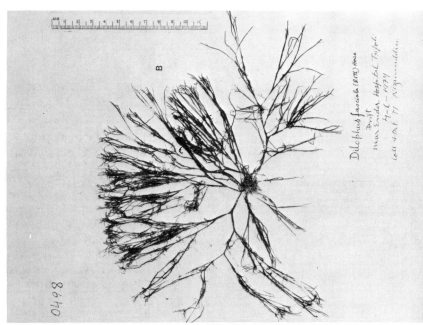

Plate XVIII. *Dictyota pusilla:* (A) Habit of the frond. *Dilophus fasciola* (Roth) Howe: (B) Habit of the frond.

98

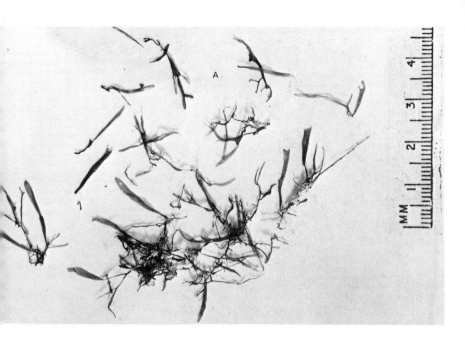

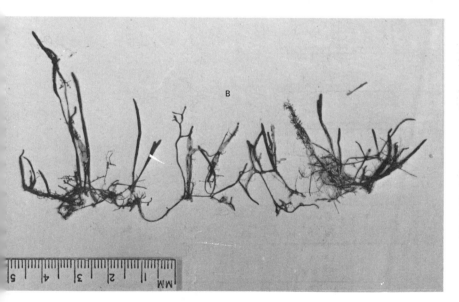

Plate XIX. *Dilophus fasciola* var. *repens* (J. Agardh) Feldmann: (A) Habit
f the frond (Black Sea, Berol. No. 23625). (B) Habit of the frond (Niece,
ectotype No. L 49523).

99

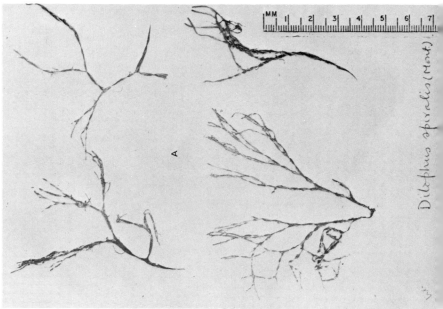

Plate XX. *Dilophus spiralis* (Montagne) Hamel: (A) Habit of the frond. (B) Habit of the frond (Endr. Hospital, Tripoli).

100

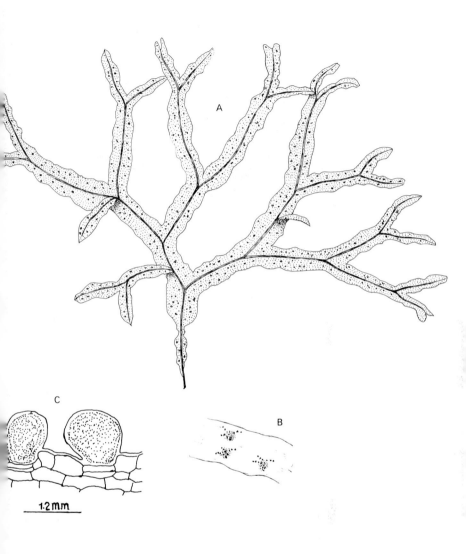

Plate XXI. *Dictyopteris membranacea:* (A) Habit of the frond (Redrawn after Stackhouse). (B) A part of a segment showing sporangia (after Stackhouse). (C) T.S. of the segment showing tetrasporangia.

101

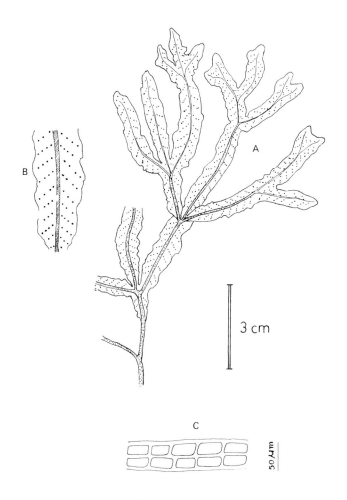

3 cm

C

50 μm

Plate XXII. *Dictyopteris tripolitana:* (A) A part of the frond. (B) Showing arrangement of sori. (C) T.S. of the frond through the margin.

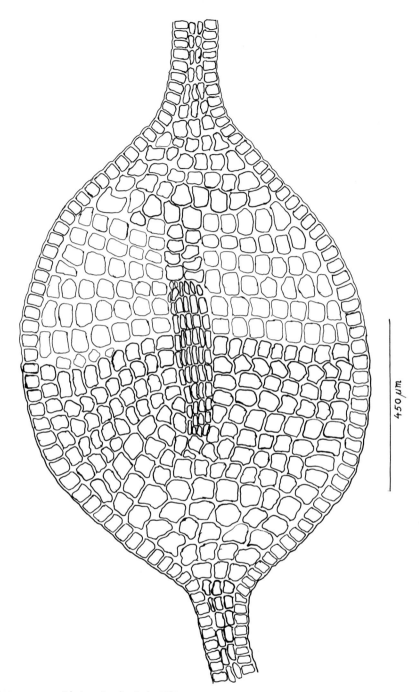

450 μm

Plate XXIII. *Dictyopteris tripolitana:* T.S. of the segment through the mid-rib.

103

Plate XXIV. *Dictyopteris polypodioides:* (A) A part of the frond. (B) A part of the segment showing arrangement of sori. (C) A part of the frond of *Ulva polypodioides* DC. in G, No. 374/5A. (D) A part of a segment showing arrangement of sori. (E) T.S. of the segment through the margin. (F) T.S. of the segment showing young antheridia. (G) T.S. of the segment showing tetrasporangia.

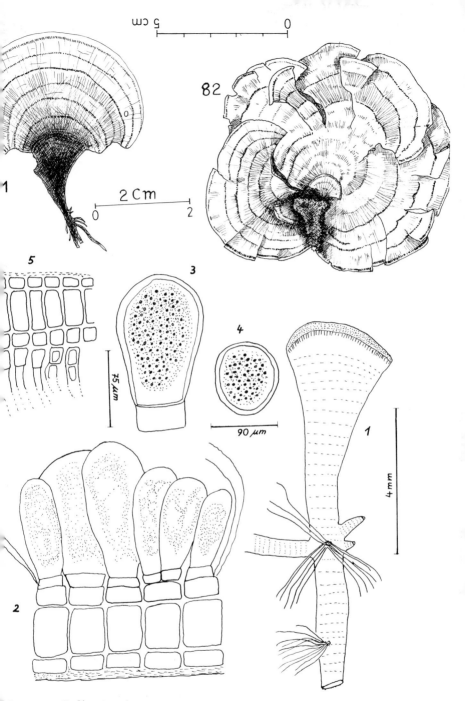

Plate XXV. *Padina tenuis:* Figs 81 & 82. Habit of the fronds showing growth phase laterals. *Padina gymnospora:* (1) Young frond with laterals and rhizoids. (2) L.S. of the frond with tetrasporangia. (3) Showing a young tetrasporangium. (4) A spore. (5) T.S. of the stipe with rhizoids.

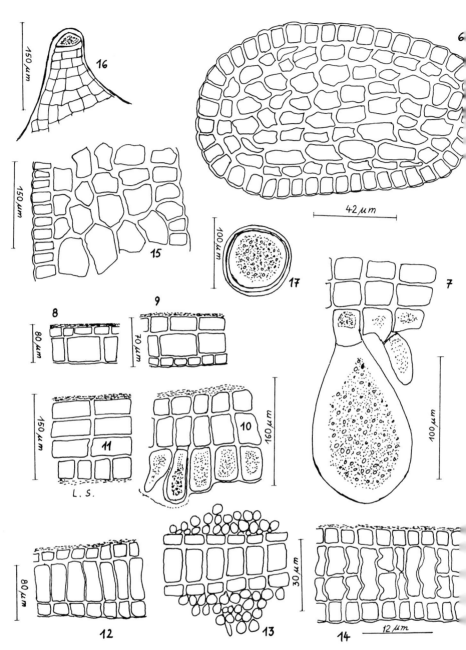

Plate XXVI. *Padina tenuis:* Fig. 6. T.S. of growth phase lateral. Fig. 7. T.
S. of the segment showing sporangia. Fig. 8. L.S. near the tip of the margin. Fig. 9. L.S. of the middle part of the frond. Fig. 10. T.S. of the
frond showing development of sporangia. Fig. 11. L.S. of the lower part of
the frond. Fig. 12. T.S. of the frond. Fig. 13. T.S. of the stipe. Fig. 14.
T.S. of the lower part of the frond. Fig. 15. T.S. of the middle part of
the lateral. Fig. 16. A part of the lateral showing an apical cell. Fig. 17.
A spore.

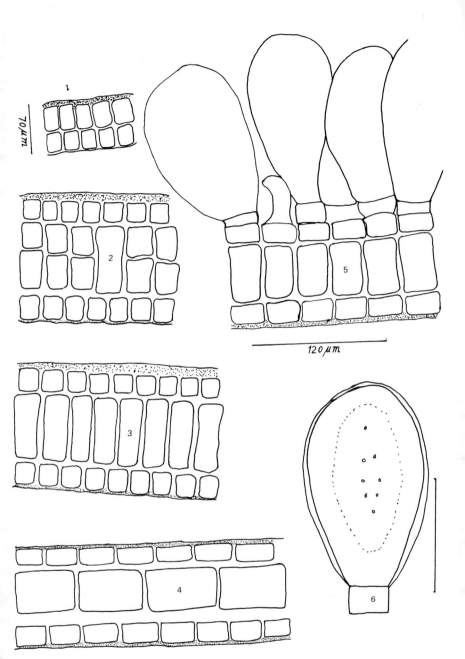

1

70 μm

2

5

120 μm

3

4

6

Plate XXVII. *Padina tenuis:* Fig. 1. T.S. through the margin of the frond.
Fig. 2. T.S. near the transition zone (between the stipe and the frond).
Fig. 3. T.S. of the stipe. Fig. 4. L.S. of the frond. Fig. 5. T.S. of the
frond showing sporangia. Fig. 6. A sporangium.

107

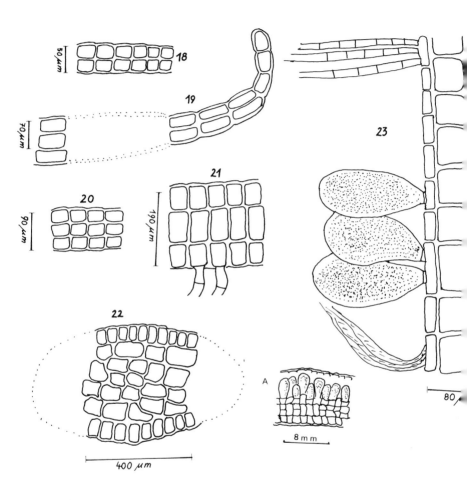

Plate XXVIII. *Padina pavonia:* Fig. 18. T.S. near the apex of the frond.
Fig. 19. L.S. near the apex. Fig. 20. T.S. of the middle part of the frond.
Fig. 21. T.S. of the lower part of the frond. Fig. 22. T.S. of the stipe.
Fig. 23. L.S. of the frond with sporangia and hairs. (A) T.S. of the frond
showing sporangia and inducium.

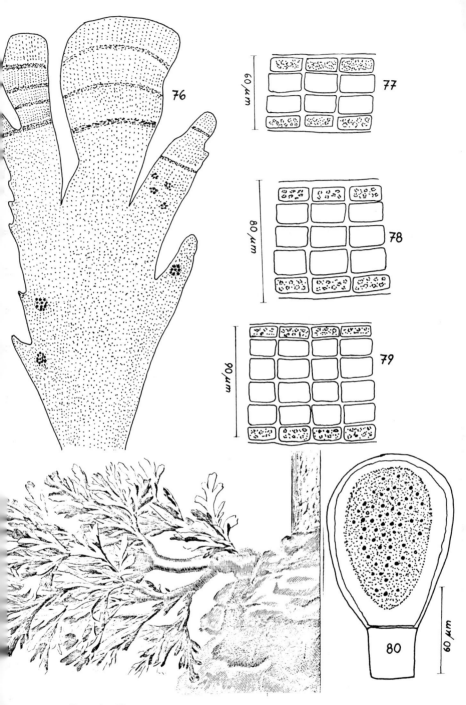

Plate XXIX. *Zonaria flava*: Fig. 76. Upper part of the frond with sporangia and zones. Fig. 77. T.S. near the apex of the frond. Fig. 78. T.S. just below the apical part of the segment. Fig. 79. T.S. through the middle part of the segment. Fig. 80. A sporangium. Tab. 336. Syn. tax. *Zonaria tournefortii* Lamx.: (Syn. tax.) *Fucus tournefortii* Lamx.

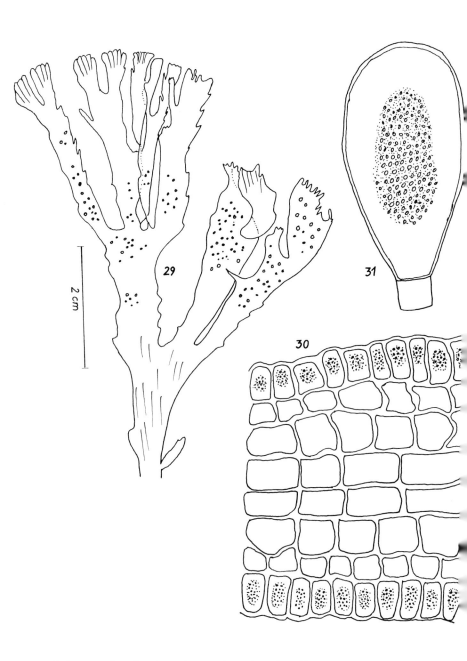

Plate XXX. *Taonia atomaria* f. *atomaria:* Fig. 29. Upper part of a segment.
Fig. 30. T.S. of the segment. Fig. 31. A sporangium.

110

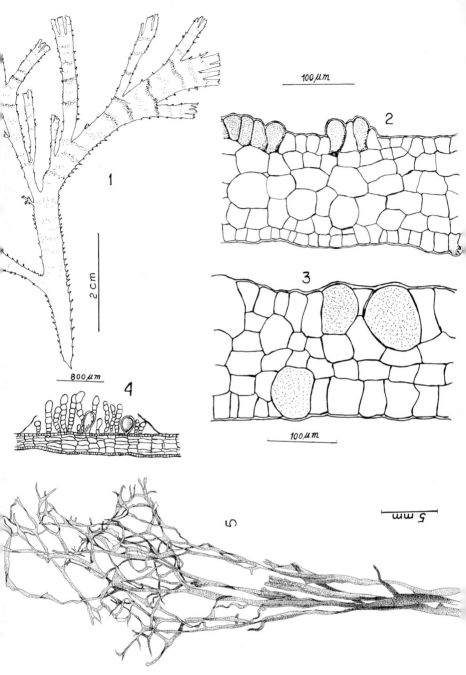

Plate XXXI. *Taonia atomaria* f. *ciliata:* Fig. 1. A part of the frond showing
dentation. *T. atomaria:* Fig. 2. T.S. of the thallus with sporangia (Redrawn
after P. Gayral, 1966). *Spatoglossum solieri:* Fig. 3. Section of the frond
with oogonium (Redrawn after P. Gayral, 1966). *Zonaria flava:* Fig. 4. T.S.
of the frond showing sporangia (Redrawn after P. Gayral, 1966). *Dictyota
dichotoma* var. *intricata:* Fig. 5. A part of lectotype *Zonaria dichotoma*
Ag.: L. No. 48878, Trieste, 1821.

Plate XXXII. *Taonia atomaria* f. *ciliata* Figs. 24-27. T.S. of the frond
showing development of sporangia. Fig. 28. A young segment of the frond
showing dentation. *Dictyota dichotoma* var. *intricata:* Fig. 36. A sporan-
dium. Fig. 37. T.S. of the segment. Fig. 38. T.S. of the segment with
oogonium. Fig. 39. T.S. of the thallus showing pit-connections. Fig. 40.
A sporangium. Fig. 41. T.S. thallus showing tetrasporangia. Fig. 73. Upper
part of the segment. Figs. 74 & 75. T.S. of the segment from two regions.

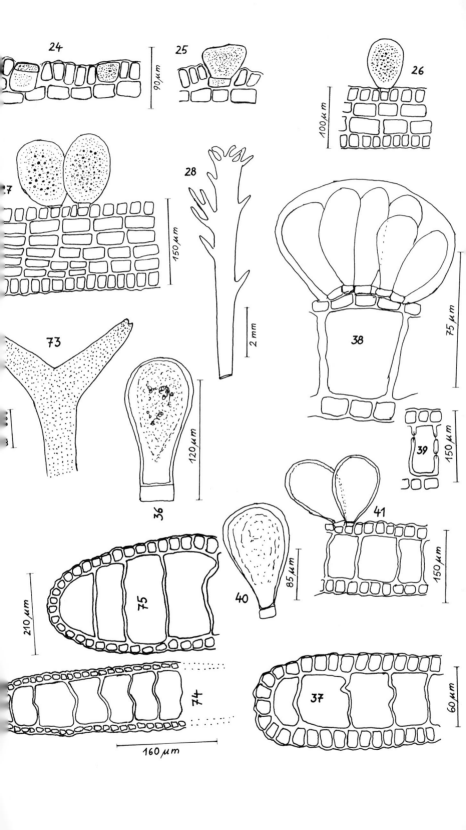

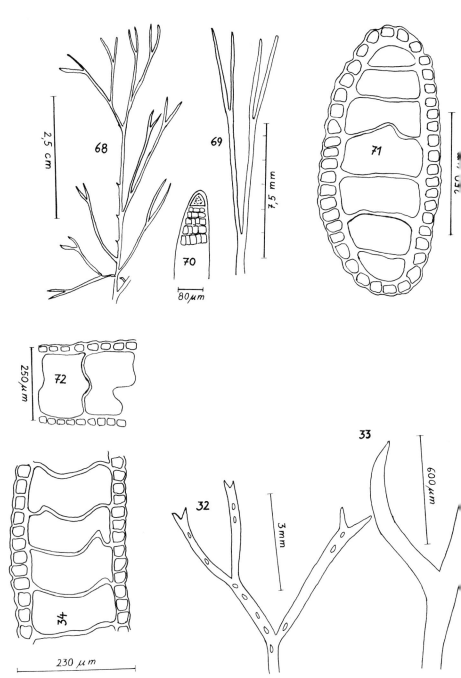

Plate XXXIII. *Dictyota linearis:* Fig. 32. Upper part of the frond. Fig. 33.
Terminal segment. Fig. 34. T.S. of the segment. *Dictyota pusilla:* Fig. 68.
Upper part of the segment. Fig. 69. A terminal segment. Fig. 70. Showing an
apical cell. Fig. 71. T.S. of the basal part of the frond. Fig. 72. T.S. of
the segment (distorted cell wall due to fixative).

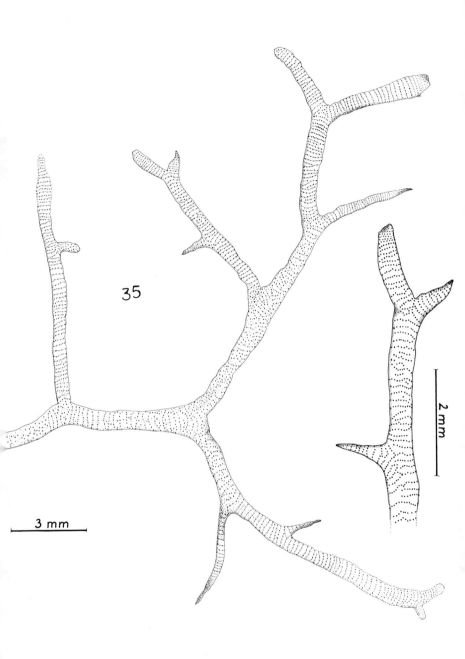

35

2 mm

3 mm

Plate XXXIV. *Dictyota linearis:* Fig. 35. Showing zones and patterns of branching.

115

Plate XXXV. *Dilophus fasciola:* Fig. 52. Terminal part of a segment with sporangia. Fig. 53. T.S. of a segment below the first dichotomy of the frond. Fig. 54. T.S. of a young proliferated lateral. Fig. 55. T.S. of the stolon. Fig. 56. T.S. of the segment with oogonia. Fig. 57. T.S. of the segment with tetrasporangia. Fig. 58. T.S. of the upper part of the segment. Fig. 59. T.S. of the basal part of the erect axis. Fig. 60. T.S. of the mature proliferated lateral. Fig. 61. A terminal part of a segment with sori. Fig. 62. Apex showing an apical cell and its segmentation.

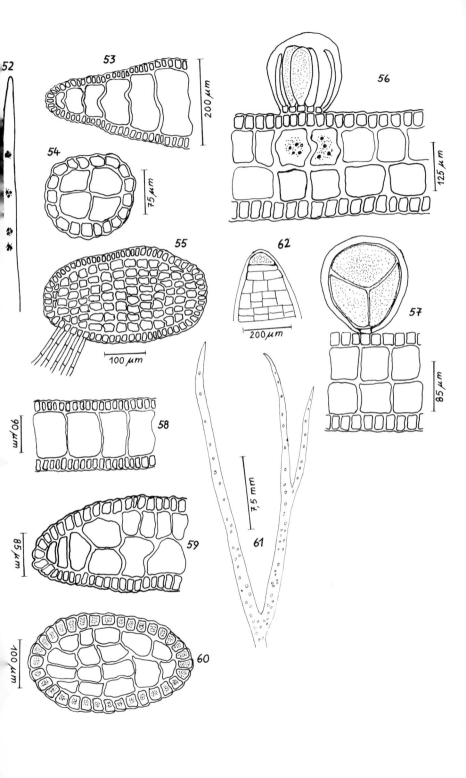

Plate XXXVI. *Dilophus spiralis:* Fig. 42. T.S. of the basal part of the erect axis. Fig. 43. T.S. of the upper part of the segment with tetra-sporangia. Fig. 44. T.S. of the stolon of the frond. Fig. 45. T.S. of the proliferated lateral. Fig. 46. T.S. of the main segment . Fig. 47. T.S. of the basal part of the lateral. *Dilophus spiralis:* Fig. 48. T.S. of a seg-ment. Fig. 49. Terminal part of the segment showing arrangement of sori. Fig. 50. Apex with apical cell and its segmentation. *Dilophus fasciola:* Fig. 51. A part of the frond.

118

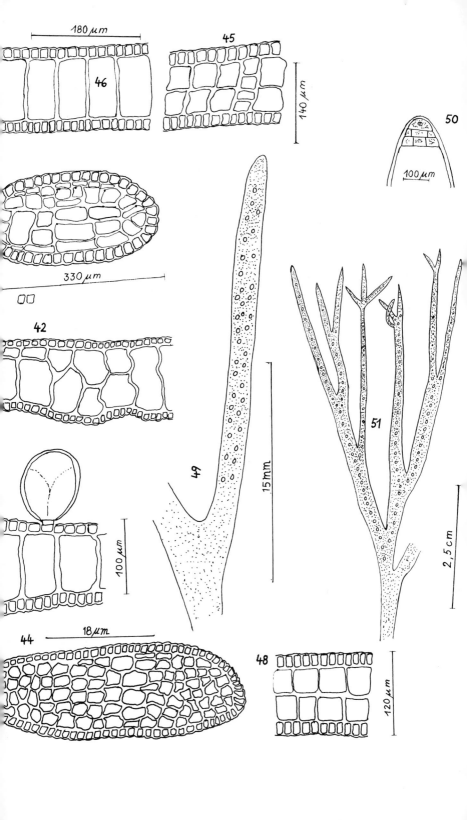

180 μm

45

46

140 μm

50

100 μm

330 μm

42

49

15 mm

51

100 μm

44

18 μm

2,5 cm

48

120 μm

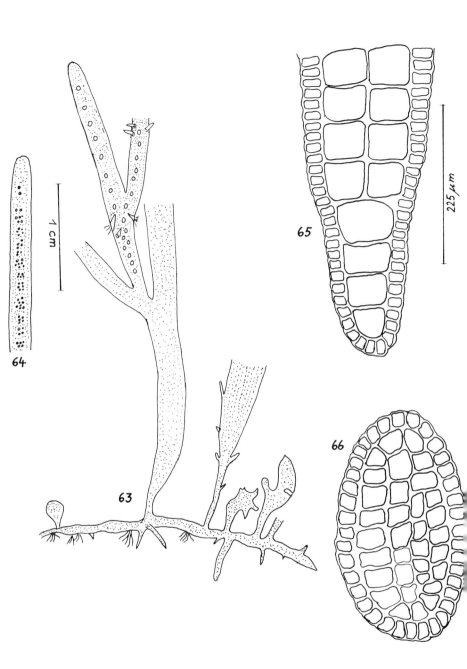

Plate XXXVII. *Dilophus spiralis:* Fig. 63. Habit of the frond. Fig. 64.
Terminal part of the segment with tetrasporangia. Fig. 65. T.S. of the
segment. Fig. 66. T.S. of the stolon.

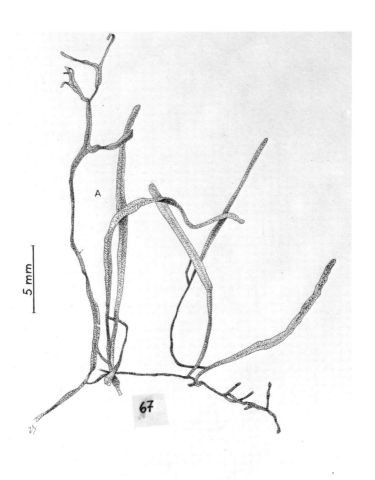

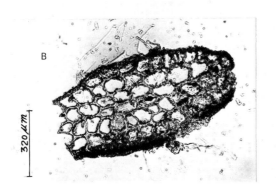

Plate XXXVIII. *Dilophus fasciola* var. *repens:* (A) Lectotype L 49523.
Habit of the frond. (B) T.S. of the frond.

121

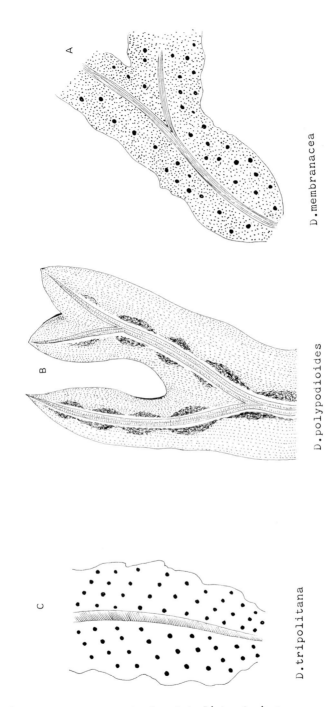

FIG. I Arrangement of sori in Dictyopteris Lamx

A — D.membranacea

B — D.polypodioides

C — D.tripolitana

Plate XXXIX. Arrangement of sori in *Dictyopteris* Lamx.

122